SpringerBriefs in Physiology

More information about this series at http://www.springer.com/series/10229

Bradley S. Fleenor • Adam J. Berrones

Arterial Stiffness

Implications and Interventions

 Springer

Bradley S. Fleenor
Department of Kinesiology and Health
  Promotion
University of Kentucky
Lexington, KY, USA

Adam J. Berrones
Department of Kinesiology and Health
  Promotion
University of Kentucky
Lexington, KY, USA

ISSN 2192-9866   ISSN 2192-9874   (electronic)
SpringerBriefs in Physiology
ISBN 978-3-319-24842-4   ISBN 978-3-319-24844-8   (eBook)
DOI 10.1007/978-3-319-24844-8

Library of Congress Control Number: 2015955254

Springer Cham Heidelberg New York Dordrecht London
© Springer International Publishing Switzerland 2015

Printed on acid-free paper

Springer International Publishing AG Switzerland is part of Springer Science+Business Media (www.springer.com)

Preface

Large elastic artery stiffness is a novel and emerging cardiovascular disease risk factor shown to predict future cardiovascular-related events and mortality. Our primary intention is to provide a concise overview for students, researchers, and clinicians by exploring the basic physiological and pathophysiological concepts and mechanisms underlying arterial stiffness. In Chap. 1, we define arterial stiffness, review the myriad ways it can be assessed, and summarize the causes, whereas in Chap. 2, we take a more in-depth look at putative cellular and molecular mechanisms of arterial stiffness. In Chap. 3, we present epidemiological factors that contribute to arterial stiffness and highlight the "bigger picture" implications for increased arterial stiffness, including the damaging effects for target organs such as the heart, brain, and kidneys. Finally, in Chap. 4, we conclude by examining clinically relevant interventions to reduce arterial stiffness that, in turn, may reduce cardiovascular disease risk and target organ damage.

Lexington, KY

Bradley S. Fleenor
Adam J. Berrones

Contents

Chapter 1
Overview of Arterial Stiffness

Abstract Arterial stiffness is a novel risk factor for cardiovascular disease resulting from functional and structural changes within arteries. Greater arterial stiffening
places negative consequences on the heart, because greater effort is required for the
myocardium to overcome the resultant increase in afterload. While the etiology of
arterial stiffness is not entirely clear, aging appears to be the principal accelerator of
the functional and structural remodeling that occurs in the arteries after the third
decade of life. Importantly, arterial stiffness can be easily assessed with potential for
routine clinical measurement.

Keywords Aorta • Stiffness • Cardiovascular • Pulse wave velocity • Mortality

What Is Arterial Stiffness?

Quoted in the seventeenth century, English physician Thomas Sydenham stated that
"A man is as old as his arteries" [1]. Indeed, a historical review of the advances in
vascular physiology has shown a shift in viewpoint of the vessels as mere passive
conduits of blood, to active contributors of cardiovascular disease (CVD). As such,
the influence of vascular pathophysiology has been increasingly studied as a major
participant in CVD-related events including heart attacks and strokes. One vascular
consequence that has been linked to increased CVD risk and events is arterial stiffness. Thus, this chapter will introduce arterial stiffness by: (1) overviewing the
physiological and pathophysiological aspects of arterial stiffness, and the contributing factors, (2) discussing commonly used non-invasive approaches for assessing
arterial stiffness, and (3) highlighting the pathophysiology of arterial stiffness. This
initial chapter will provide the basis and foundation for the fundamental principles
of arterial stiffness and potential consequences that will be examined in more detail
in the subsequent chapters.

   Because there is potential for confusion, we will address the differences in **arteriosclerosis** versus **atherosclerosis** to delimit the scope of this monograph.
Arteriosclerosis is the thickening and loss of elasticity of the arterial wall that
occurs with normal, healthy aging, and in many disease conditions including atherosclerosis [2]. **Atherosclerosis**, however, is characterized by localized and
obstructive plaque buildup in the arterial wall, which is a common condition leading

B.S. Fleenor, A.J. Berrones, *Arterial Stiffness*, SpringerBriefs in Physiology,
DOI 10.1007/978-3-319-24844-8_1

to **arteriosclerosis**. In short, arteriosclerosis can occur in the absence or presence of overt disease such as atherosclerosis. Due to the extensive study of arteriosclerosis with older age, much of the discussion herein will be based on data from aging models. We will, however, highlight other disease conditions where arterial stiffening emerges as an important CVD risk factor.

Pathophysiology of Arterial Stiffness

Aortic Function and Physiology

To better understand the pathophysiology of arterial stiffness, it is important to briefly review the functional role and physiology of the arterial system. The primary function of the arterial system is to provide oxygen and nutrient rich blood to the various tissues and organs of the body. In addition, the arterial system, specifically the thoracic aorta, buffers the blood ejected from the heart by extending and recoiling to promote continuous blood flow in the capillaries [3]. This design has been likened to that of a windkessel of antique fire engines [4], which is purposeful for cushioning the left ventricular blood ejected during systole, and for continuous propagation of blood flow through the arterial tree during diastole.

Each contraction of the heart results in a pulse pressure wave that is emitted from the heart, which precedes blood flow down the aorta. Under normal, healthy physiological conditions the pulse pressure wave travels down the aorta and is partially reflected back to the heart while the remaining wave is transmitted to the microcirculation to promote low pulsatile capillary blood flow in tissues. The return, or reflective wave arrives at the heart during diastole, which provides the driving pressure for coronary artery perfusion during diastole. When the aorta is stiffened, as observed with aging, the pulse pressure wave travels down the aorta at an increased speed with a greater proportion of the wave being transmitted to the microcirculation. The increased transmitted wave results in greater pulsatile blood flow in tissue capillaries, which causes damage to these small vessels that, in turn, may contribute to target (end) organ damage of the heart, brain, and kidneys (Fig. 1.1) [5, 6].

In addition, because the pulse wave travels faster through a stiffened aorta, the reflected wave returns to the heart sooner, during late systole when the heart is ejecting blood. This results in an increase in afterload, or greater resistance by which the left ventricle must overcome to pump blood from the heart, and thus is a contributing factor to hypertension. Notably, the early arrival of the pulse pressure wave to the heart also decreases the perfusion pressure in the coronary arteries, which may reduce coronary blood flow at rest and/or during physiological stress such as exercise. Thus, increased arterial stiffness—specifically aortic stiffness—is implicated in heart-related pathologies including hypertension and myocardial ischemia (Fig. 1.1).

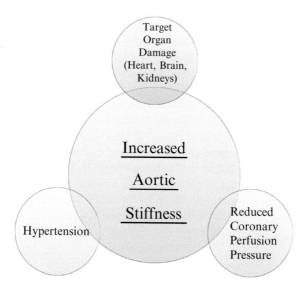

Fig. 1.1 Arterial stiffness is an important target for a variety of pathological conditions

Implications Throughout the Arterial Tree

In general, the aging process accelerates vessel wall stiffening, which is more pronounced in the large elastic arteries (aorta, carotid) than in the peripheral arteries (brachial, tibial). As such, aging from early to mid-life (10–50 years of age) results an approximately 70 % increase in aortic stiffness, whereas peripheral artery stiffness increases only ~20 % [7]. Because arterial stiffness is non-uniform in its development, and appears to be enhanced in the large elastic arteries, the impact to cardiovascular health will vary throughout systemic circulation in a site-specific manner. We will discuss the differences between the large elastic versus the medium and small arteries.

Large Elastic Arteries

By convention, discussion of the macro-vasculature involves arteries with an internal diameter greater than 10 mm, up to the largest elastic artery in the human body, the proximal aorta (25 mm luminal diameter) [4]. The principal large elastic artery is the aorta, which has a diverse function and structural composition throughout. The thoracic aorta, from the heart to the diaphragm, is the most elastic segment containing a greater proportion of elastin compared with collagen. The greater elastin composition allows this segment of aorta to be more distensible to buffer the bolus of blood that is ejected from the heart that, in turn, recoils to propel the blood down the aorta during diastole. The distal aorta, which spans from the diaphragm to the iliac artery promotes and distributes oxygen and nutrient dense blood to tissues and has greater amounts of the rigid protein collagen, and proportionally less elastin.

Importantly, the stiffness of arteries increases as a function of distance from the heart in order to perform unique functions. Because of this biological arrangement, the proximal aorta has a greater potential to stiffen due in part to the greater stress placed on this aortic segment to buffer the ejected blood with each heart beat compared with the distal aorta or smaller arteries [8]. As such, a logarithmic relationship between linear strain (or stretch) and the number of cycles determines the extent of fatigue and fracture in the elastin and collagen fibers [7]. Thus, by holding constant the number of cardiac cycles, we see that aortic stiffness is increased (experimentally) as a function of the magnitude of linear strain (stretch), which implicates blood pressure as an important modulator of arterial stiffness.

Muscular (Small and Medium) Arteries and Arterioles

Medium and small arteries have a luminal diameter of 0.50–10 mm [9] with a decreased elastin to collagen ratio making them stiffer compared with large arteries. Medium sized arteries help regulate blood flow in various tissues via local vasodilation and vasoconstriction. Small arteries and arterioles (<0.50 mm luminal diameter) have greater frictional resistance to blood flow compared with large and medium sized arteries. Thus, the higher frictional resistance and cross-sectional area of small arteries and arterioles control blood pressure, and ultimately determine mean arterial pressure [4]. The small muscular arteries, in essence, function as stopcocks of the vascular system and are responsible for the steep reduction in mean arterial pressure across the vasculature. However, it appears, that the medium and small arteries do not become stiffer despite increased stiffening of large elastic arteries such as the carotid and aorta with aging or hypertension [10–12]. Thus, these findings collectively indicate an effect of aging and disease to increase central large artery stiffness, specifically the aorta and carotid arteries, compared with the small and medium arteries of the peripheral vasculature.

Aspects of Arterial Stiffness

Arterial stiffness is associated with both structural and functional changes within the artery. Impaired smooth muscle function resulting in altered arterial dilation and constriction, and increased blood pressure is an important functional mechanism for increased arterial stiffness. Additionally, the abundance and organization of the extracellular matrix composition also contribute to the alterations of arterial structure. These functional and structural changes within arteries will be briefly summarized.

Functional Changes

Arterial stiffness in older adults is associated with isolated systolic hypertension and greater CVD risk [13]. Because aging is irreversible, but blood pressure is (to an extent) modifiable, recognition and treatment of isolated systolic hypertension is an important target to attenuate the pathophysiological effects of arterial stiffness. However, because of the close relationship between systolic blood pressure (SBP) and aortic stiffness, there is some question as to whether arterial stiffness leads to hypertension, or hypertension promotes arterial stiffness [14]. Regardless of underlying pathological process, reducing blood pressure to decrease arterial stiffness is an important therapeutic target. As such, impaired smooth muscle function leading to greater aortic tone, is one target to decrease systolic blood pressure. It has recently been postulated that nearly 50 % of age-related aortic stiffening is due to aortic vascular smooth muscle cells (VSMCs) [15]. Importantly, arterial stiffness as a result of impaired VSMC function is, in part, attributable to endothelial dysfunction and reduced nitric oxide bioavailability [16–19]. Therefore, reducing blood pressure is a key target for decreasing aortic stiffness, which may be accomplished by improving both VSMC function and nitric oxide bioavailability.

Structural Changes

In addition to the functional regulation of blood pressure, structural alterations within vessels also contribute to arterial stiffening [20]. These changes include thickening and remodeling within each of the three layers of the artery, which include the: tunica intima, tunica media, and tunica adventitia (Fig. 1.2) [21]. While the tunica intima and media have been studied more extensively, the outermost, less-studied tunica adventitia layer has also been shown to remodel in conditions of increased arterial stiffening [7]. Each layer of the artery either has an increase in thickness and is geometrically remodeled, or has demonstrated structural remodeling of extracellular matrix (ECM) proteins, both of which promote arterial stiffness.

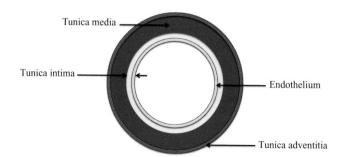

Fig. 1.2 The structural layers of an artery

Structural remodeling of the ECM is related, in large part, to increased collagen I deposition, a key load-bearing collagen isoform in arteries, and their cross-linking by advanced glycation end products (AGEs), which are also in greater abundance of stiffer arteries. In contrast, elastin, a protein that provides elasticity to arteries has an attenuated expression that further contributes to aortic stiffening. Moreover, a disordered and/or fragmented arrangement of the ECM also promotes arterial stiffness [22]. There are many unique extracellular-related changes within arteries contributing to arterial stiffness. For example, cellular contributions to arterial stiffness such as stiffening of individual vascular smooth muscle cells (VSMCs) with aging are related to specific cytoskeletal compositions within the cell [23]. Although only limited information is available, the cell-specific role of VSMCs is an important and emerging mechanistic area of investigation for arterial stiffness-related pathologies and warrants further investigation.

Assessing Arterial Stiffness in Adults

The changes within arteries leading to age-related aortic stiffening are largely due to functional and structural changes within the artery [7]. These arterial changes result in volume and pressure changes, which can be used to assess arterial stiffness. For example, assessing the compliance, or elasticity of an artery can be determined by quantifying the changes in volume and pressure (compliance = $\Delta V/\Delta P$) during a cardiac cycle. Pulse pressure (systolic minus diastolic blood pressure) can also be easily assessed clinically and used as a surrogate measure of arterial stiffness. To gain even greater insight, however, Young's modulus, a ratio of force per unit area (stress, units = Pa) per unit length (strain, ratio of the deformation to its original form, units = dimensionless), has also been used to characterize arterial wall stiffness using the incremental elastic modulus (E_{inc}). The E_{inc} can then be used to calculate pulse wave velocity (PWV) as described in the Moens-Kortweg equation: $PWV = \sqrt{(E_{inc} \cdot h/2\ r\ \rho)}$, where h is the vessel wall thickness, r is the vessel radius, and ρ is the density of blood. Thus, age-related aortic stiffness increases the PWV, which can readily be assessed across the arterial tree [7].

Pulse Wave Velocity

The viscoelastic properties of arteries vary widely across the arterial tree. Thus, it becomes clear that extrapolating segmental arterial properties to the whole arterial tree may not always be justified [24]. In that regard, however, local or regional specific measures of arterial stiffness of the aorta or carotid arteries can be assessed non-invasively, using ultrasonic echotracking systems, magnetic resonance imaging (MRI), or by using the pulse pressure wave that is generated with each contraction of the heart [24]. The velocity of the pressure wave also varies across the arterial tree, such that, the velocity of this pressure wave is lower in the proximal aorta than in the periphery indicating the aorta is less stiff than peripheral arteries [4]. Consequently, when arterial stiffness is high, as in aging, so is the PWV. The pressure wave can be conveniently palpated at the radial, femoral, or carotid arteries for routine and repeatable measures.

To determine regional arterial stiffness of the aorta, which includes the entire aorta except the ascending segment, non-invasive carotid-femoral PWV (cfPWV) can be utilized and is considered the 'gold-standard' measurement [24]. To obtain this measure the pulse waveforms are acquired non-invasively, either sequentially or simultaneously, at the carotid and femoral arteries using tonometry or Doppler flow probes. The superficial distance (**D**, meters) between the right common carotid and right femoral arteries, and the time delay (**Δt** or transit time, seconds) between the feet of the two respective waveforms are acquired. By measuring the distance between the carotid and femoral arteries relative to the suprasternal notch, a reference point, height is accounted for in this measurement. cfPWV is therefore calculated as D (meters)/Δt (seconds), which provides a regional measure of aortic stiffness (Fig. 1.3). Logically, the greater the velocity of the pulse wave traveling down the aorta the greater the extent of arterial stiffening. Importantly, this 'gold-standard' measure of arterial stiffness is associated with cardiovascular events, independently of conventional risk factors [3].

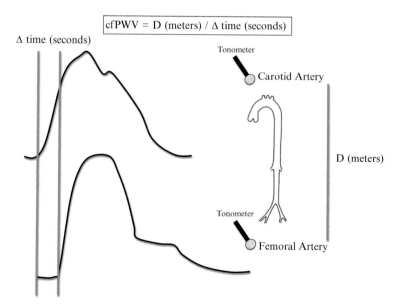

Fig. 1.3 The pulse wave transit time provides a regional measure of aortic stiffness. Time is calculated from the foot of the carotid and femoral waveforms, and superficial distance is measured from the carotid to femoral sites

Pulse Wave Analysis to Assess Central Hemodynamic Parameters

Applanation tonometry is a technique used to assess central hemodynamic parameters that can influence aortic stiffness (Fig. 1.4). There are many commercially available devices, but the SphygmoCor system (AtCor Medical, West Ryde, Australia) is one of the most widely used devices. Importantly, these noninvasive estimations of central blood pressure correlate well with invasive catheter-derived measurements acquired at the ascending aorta with a correlation coefficient of 0.91 (demonstrated with the SphygmoCor device) [25]. The SphygmoCor applanation device uses a generalized mathematical transfer function to create an aortic pressure waveform that is computed both from the pressure waveform at the radial artery, as well as the brachial blood pressure. The Food and Drug Administration has accepted the central blood pressure values derived from peripheral waveforms via applanation tonometry as a "substantially equivalent" assessment of aortic pressure when compared with invasive catheterization measures [3]. Hence, applanation tonometry is a validated, non-invasive alternative to catheterization and is commonly used in research settings.

Central Pulse Pressure

Pulse pressure, or the difference between systolic and diastolic blood pressure within arteries is an important surrogate marker for arterial stiffness. Based on the pressure-strain elastic modulus, E_p to define arterial stiffness, which is calculated by the equation $E_p = \Delta P/(\Delta D/D)$, where ΔP is the aortic pulse pressure; ΔD is the maximal change in aortic diameter during the cardiac cycle; and D is the mean aortic diameter during the cardiac cycle. This characterization of arterial stiffness points to

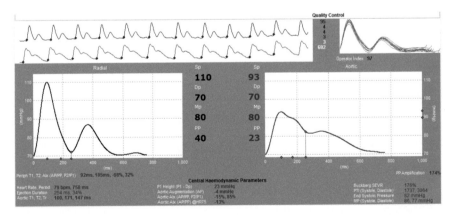

Fig. 1.4 Screen shot of Pulse Wave Analysis and central hemodynamic parameters using the SphygmoCor AtCor Medical v. 9.0

elevated central pulse pressure as a major participant in the disease's etiology. Ultimately, the pressure-strain elastic modulus will be elevated in individuals whose aorta is unable to increase luminal diameter for a given pressure. Hence, pulse pressure is the pulsatile component of blood pressure that drives repetitive strain and contributes to fragmentation of aortic elastin, which can be used as a surrogate marker of arterial stiffness [14]. It is important to consider within the context of enhancing cardiovascular risk prediction, that the routine use of central pulse pressure measurements in clinical practice has not been substantiated [26].

Augmentation Index (AIx)

AIx is the difference between the first and second systolic peaks relative to pulse pressure [24]. The AIx measure is based on the return, or reflected, wave that is transmitted with each heartbeat. In arteries that confer greater stiffness the reflected wave returns to the left ventricle of the heart sooner causing amplification of the systolic blood pressure. The early arrival of the reflected wave increases the myocardial oxygen demand and requires greater mechanical work to overcome the augmented aortic root pressure (afterload), which is one deleterious consequence of arterial stiffness. AIx has also been used as a surrogate measure of arterial stiffness [27].

    While the interpretation of AIx is straightforward, one limitation of using this measure as a substitute of cfPWV is AIx has a nonlinear relation with arterial stiffness with increasing age [28]. After approximately 60 years of age AIx drops, but arterial stiffness when measured by the cfPWV, increases [29, 30]. Therefore AIx is considered to be a more global, or whole-body, measure of stiffness, rather than a specific aortic stiffening endpoint as it assesses the cumulative reflected waves at the heart. Thus, AIx may be used to describe the magnitude of peripheral wave reflections, but care should be taken when using AIx as a measure of arterial stiffness, particularly in older populations.

Intima-Media Thickness (IMT)

Measuring IMT with ultrasound devices has been used as a surrogate for determining local arterial stiffness, and is typically assessed in the carotid artery [24]. One potential limitation with the ultrasound-derived measure of IMT is that the calculation of Young's modulus assumes homogeneity of the arterial wall, but—by design—the arteries are inhomogeneous [8]. Further, the intima cannot be distinguished from the media with ultrasound [8], and hyperplasia of the intima, compared with the medial layer, is the major cause of increased IMT [31]. As a result, aortic morphology is dependent on age, and so calculation of Young's modulus from the IMT will yield uncertain measures of arterial stiffness.

For example, prospective data from the Multi-Ethnic Study of Atherosclerosis (MESA) study showed arterial calcification and plaque presence were stronger predictors of incident CVD versus increased IMT alone [32]. And, while IMT does correlate with aPWV in normal subjects, IMT does not appear to be an interchangeable measure of arterial stiffness in high-risk patients such as those with hypertension and/or diabetes [24]. Hence, in cases of high-risk patients, the rate of aortic stiffening relative to carotid IMT thickening is increased, but is unpredictable when the carotid and aorta do not stiffen at the same rate [33]. More prospective trials are needed to qualify the use of IMT as a sole determinant of arterial stiffness.

Pathophysiological Role of Arterial Stiffness in CVD

Traditional measures of blood pressure occur in the brachial artery, a relatively muscular peripheral artery. However, the hemodynamics, or blood pressure, in the aorta is markedly different from what is observed in the brachial artery. One example of the discrepancy between the central and peripheral vasculature is the amplification phenomenon, which shows that pulse pressure is lower in the aorta compared with brachial artery. However, in conditions of aging and/or disease pulse pressure amplification is decreased due to increases in aortic stiffness and blood pressure. In addition, central hemodynamic parameters such as pulse pressure can be accurately and easily obtained with tonometry; additionally, quartiles of central pulse pressure are stronger predictors of future CV events in contrast to quartiles of peripheral pulse pressure [34]. Clinicians have historically relied on and continue to use peripheral blood pressure measures to determine treatment options. However, current data suggest treatments based on central blood pressures may be more efficacious [35–37].

Increase in SBP, Central Pulse Pressure and Left Ventricle Afterload

Arterial stiffness accelerates the return waves back to the left ventricle of heart, and elevates SBP and pulse pressure. The paradoxical question, however, is: Does hypertension cause arterial stiffness or does arterial stiffness cause hypertension. This relationship has been summarized as the colloquial chicken-and-egg dilemma [14]. The current general viewpoint is that hypertension in young adults accelerates arterial stiffness, similar to that of aging, owing to the premature return of reflected waves in late systole [24]. Evidence also indicates arterial stiffness precedes the rise in blood pressure [14]. Regardless, an increase in SBP results in the heart working harder to overcome the increased pressure in the aortic root (afterload). Hence, arterial stiffness is associated with left ventricular hypertrophy (LVH) [38, 39], which is an established risk factor for CVD [7, 32, 39, 40].

Decrease in Coronary Perfusion Pressure

Isolated systolic hypertension is characterized by a normal diastolic blood pressure (DBP), but with a systolic pressure greater than 140 mmHg. Because the vast majority of the myocardial blood flow occurs during diastole, sufficient coronary perfusion pressure is required to oxygenate the heart. Interestingly, DBP increases with age up until approximately 60 years, and then decreases [9]. Remember that arterial stiffness increases dramatically after 60 years of age, and due to elevated SBP, the fall in DBP with increasing arterial stiffness is explained by a diminished hydraulic buffering system leading to greater peripheral run-off of stroke volume during systole [9]. Results from the Framingham study indicate that if hypertension is left untreated, then elevated DBP is associated with increased SBP, allowing arterial stiffness to perpetuate in an accelerated and vicious circle [41]. Collectively, these findings indicate that greater aortic stiffness in older age reduces coronary perfusion pressure, which is possibly linked to myocardial ischemia.

Summary

Arterial stiffness, or arteriosclerosis, is an emerging independent risk factor for the prediction of future CVD-related events. Non-invasive approaches for assessing arterial stiffness make this measure readily available to clinicians for routine assessment. The principal causes of arterial stiffness are the functional and structural changes that accompany aging with the greatest changes observed in large elastic arteries (aorta, carotid) compared to peripheral arteries (brachial, tibial). Importantly, increased aortic stiffness is a novel risk factor associated with pathological conditions of the heart, brain, and kidneys with potential for routine clinical use.

References

1. National Institute on Aging (2005) Aging hearts & arteries: a scientific quest. National Institute of Health, National Institute on Aging, U.S. Department of Health and Human Services, Bethesda
2. (1988) Dorland's illustrated medical dictionary, 27 edn. WB Saunders, Philadelphia
3. O'Rourke MF, Hashimoto J (2007) Mechanical factors in arterial aging: a clinical perspective. J Am Coll Cardiol 50(1):1–13. doi:10.1016/j.jacc.2006.12.050
4. Levy MN, Pappano AJ (2007) Cardiovascular physiology, 9th edn. Mosby, Philadelphia
5. Mitchell GF (2015) Arterial stiffness: insights from Framingham and Iceland. Curr Opin Nephrol Hypertens 24(1):1–7. doi:10.1097/mnh.0000000000000092
6. Safar ME, Nilsson PM, Blacher J, Mimran A (2012) Pulse pressure, arterial stiffness, and end-organ damage. Curr Hypertens Rep 14(4):339–344. doi:10.1007/s11906-012-0272-9
7. Nichols WW, O'Rourke MF (1990) McDonald's blood flow in arteries: theoretical, experimental, and clinical principles, 3rd edn. Edward Arnold, Philadelphia

8. Adji A, O'Rourke MF, Namasivayam M (2011) Arterial stiffness, its assessment, prognostic value, and implications for treatment. Am J Hypertens 24(1):5–17. doi:10.1038/ajh.2010.192

9. Safar ME, O'Rourke MF (2006) Arterial stiffness in hypertension, 1st edn. Elsevier, Amsterdam

10. Bortolotto LA, Hanon O, Franconi G, Boutouyrie P, Legrain S, Girerd X (1999) The aging process modifies the distensibility of elastic but not muscular arteries. Hypertension 34(4 pt 2):889–892

11. Laurent S, Girerd X, Mourad JJ, Lacolley P, Beck L, Boutouyrie P, Mignot JP, Safar M (1994) Elastic modulus of the radial artery wall material is not increased in patients with essential hypertension. Arterioscler Thromb 14(7):1223–1231

12. Zhang Y, Agnoletti D, Protogerou AD, Topouchian J, Wang JG, Xu Y, Blacher J, Safar ME (2013) Characteristics of pulse wave velocity in elastic and muscular arteries: a mismatch beyond age. J Hypertens 31(3):554–559. doi:10.1097/HJH.0b013e32835d4aec, discussion 559

13. Mackenzie IS, McEniery CM, Dhakam Z, Brown MJ, Cockcroft JR, Wilkinson IB (2009) Comparison of the effects of antihypertensive agents on central blood pressure and arterial stiffness in isolated systolic hypertension. Hypertension 54(2):409–413. doi:10.1161/hypertensionaha.109.133801

14. Mitchell GF (2014) Arterial stiffness and hypertension: chicken or egg? Hypertension. doi:10.1161/hypertensionaha.114.03449

15. Gao YZ, Saphirstein RJ, Yamin R, Suki B, Morgan KG (2014) Aging impairs smooth muscle-mediated regulation of aortic stiffness: a defect in shock absorption function? Am J Physiol Heart Circ Physiol 307(8):H1252–H1261. doi:10.1152/ajpheart.00392.2014

16. Fleenor BS, Seals DR, Zigler ML, Sindler AL (2012) Superoxide-lowering therapy with TEMPOL reverses arterial dysfunction with aging in mice. Aging Cell 11(2):269–276. doi:10.1111/j.1474-9726.2011.00783.x

17. Fleenor BS, Sindler AL, Marvi NK, Howell KL, Zigler ML, Yoshizawa M, Seals DR (2013) Curcumin ameliorates arterial dysfunction and oxidative stress with aging. Exp Gerontol 48(2):269–276. doi:10.1016/j.exger.2012.10.008

18. Kim JH, Bugaj LJ, Oh YJ, Bivalacqua TJ, Ryoo S, Soucy KG, Santhanam L, Webb A, Camara A, Sikka G, Nyhan D, Shoukas AA, Ilies M, Christianson DW, Champion HC, Berkowitz DE (2009) Arginase inhibition restores NOS coupling and reverses endothelial dysfunction and vascular stiffness in old rats. J Appl Physiol 107(4):1249–1257. doi:10.1152/japplphysiol.91393.2008

19. Soucy KG, Ryoo S, Benjo A, Lim HK, Gupta G, Sohi JS, Elser J, Aon MA, Nyhan D, Shoukas AA, Berkowitz DE (2006) Impaired shear stress-induced nitric oxide production through decreased NOS phosphorylation contributes to age-related vascular stiffness. J Appl Physiol 101(6):1751–1759. doi:10.1152/japplphysiol.00138.2006

20. Fleenor BS, Marshall KD, Durrant JR, Lesniewski LA, Seals DR (2010) Arterial stiffening with ageing is associated with transforming growth factor-beta1-related changes in adventitial collagen: reversal by aerobic exercise. J Physiol 588(pt 20):3971–3982. doi:10.1113/jphysiol.2010.194753

21. Fleenor BS (2013) Large elastic artery stiffness with aging: novel translational mechanisms and interventions. Aging Dis 4(2):76–83

22. Dodson RB, Rozance PJ, Fleenor BS, Petrash CC, Shoemaker LG, Hunter KS, Ferguson VL (2013) Increased arterial stiffness and extracellular matrix reorganization in intrauterine growth-restricted fetal sheep. Pediatr Res 73(2):147–154. doi:10.1038/pr.2012.156

23. Qiu H, Zhu Y, Sun Z, Trzeciakowski JP, Gansner M, Depre C, Resuello RR, Natividad FF, Hunter WC, Genin GM, Elson EL, Vatner DE, Meininger GA, Vatner SF (2010) Short communication: vascular smooth muscle cell stiffness as a mechanism for increased aortic stiffness with aging. Circ Res 107(5):615–619. doi:10.1161/circresaha.110.221846

24. Laurent S, Cockcroft J, Van Bortel L, Boutouyrie P, Giannattasio C, Hayoz D, Pannier B, Vlachopoulos C, Wilkinson I, Struijker-Boudier H (2006) Expert consensus document on arterial stiffness: methodological issues and clinical applications. Eur Heart J 27(21):2588–2605. doi:10.1093/eurheartj/ehl254

25. Ding FH, Fan WX, Zhang RY, Zhang Q, Li Y, Wang JG (2011) Validation of the noninvasive assessment of central blood pressure by the SphygmoCor and Omron devices against the invasive catheter measurement. Am J Hypertens 24(12):1306–1311. doi:10.1038/ajh.2011.145

26. Mitchell GF (2015) Central pressure should not be used in clinical practice. Artery Res 9:8–13. doi:10.1016/j.artres.2014.11.002

27. Mitchell GF (2009) Arterial stiffness and wave reflection: biomarkers of cardiovascular risk. Artery Res 3(2):56–64. doi:10.1016/j.artres.2009.02.002

28. Westerhof BE, Westerhof N (2012) Magnitude and return time of the reflected wave: the effects of large artery stiffness and aortic geometry. J Hypertens 30(5):932–939. doi:10.1097/HJH.0b013e3283524932

29. Mitchell GF, Parise H, Benjamin EJ, Larson MG, Keyes MJ, Vita JA, Vasan RS, Levy D (2004) Changes in arterial stiffness and wave reflection with advancing age in healthy men and women: the Framingham Heart Study. Hypertension 43(6):1239–1245. doi:10.1161/01.HYP.0000128420.01881.aa

30. McEniery CM, Yasmin HIR, Qasem A, Wilkinson IB, Cockcroft JR (2005) Normal vascular aging: differential effects on wave reflection and aortic pulse wave velocity: the Anglo-Cardiff Collaborative Trial (ACCT). J Am Coll Cardiol 46(9):1753–1760. doi:10.1016/j.jacc.2005.07.037

31. Virmani R, Avolio AP, Mergner WJ, Robinowitz M, Herderick EE, Cornhill JF, Guo SY, Liu TH, Ou DY, O'Rourke M (1991) Effect of aging on aortic morphology in populations with high and low prevalence of hypertension and atherosclerosis. Comparison between occidental and Chinese communities. Am J Pathol 139(5):1119–1129

32. Gepner AD, Young R, Delaney JA, Tattersall MC, Blaha MJ, Post WS, Gottesman RF, Kronmal R, Budoff MJ, Burke GL, Folsom AR, Liu K, Kaufman J, Stein JH (2015) Comparison of coronary artery calcium presence, carotid plaque presence, and carotid intima-media thickness for cardiovascular disease prediction in the multi-ethnic study of atherosclerosis. Circ Cardiovasc Imaging 8(1). doi:10.1161/circimaging.114.002262

33. Paini A, Boutouyrie P, Calvet D, Tropeano AI, Laloux B, Laurent S (2006) Carotid and aortic stiffness: determinants of discrepancies. Hypertension 47(3):371–376. doi:10.1161/01.hyp.0000202052.25238.68

34. Roman MJ, Devereux RB, Kizer JR, Okin PM, Lee ET, Wang W, Umans JG, Calhoun D, Howard BV (2009) High central pulse pressure is independently associated with adverse cardiovascular outcome the strong heart study. J Am Coll Cardiol 54(18):1730–1734. doi:10.1016/j.jacc.2009.05.070

35. Borlaug BA, Olson TP, Abdelmoneim SS, Melenovsky V, Sorrell VL, Noonan K, Lin G, Redfield MM (2014) A randomized pilot study of aortic waveform guided therapy in chronic heart failure. J Am Heart Assoc 3(2), e000745. doi:10.1161/jaha.113.000745

36. Drawz PE, Abdalla M, Rahman M (2012) Blood pressure measurement: clinic, home, ambulatory, and beyond. Am J Kidney Dis 60(3):449–462. doi:10.1053/j.ajkd.2012.01.026

37. Shimizu M, Hoshide S, Ishikawa J, Yano Y, Eguchi K, Kario K (2014) Correlation of central blood pressure to hypertensive target organ damages during antihypertensive treatment: the J-TOP study. Am J Hypertens. doi:10.1093/ajh/hpu250

38. London GM, Pannier B, Guerin AP, Marchais SJ, Safar ME, Cuche JL (1994) Cardiac hypertrophy, aortic compliance, peripheral resistance, and wave reflection in end-stage renal disease. Comparative effects of ACE inhibition and calcium channel blockade. Circulation 90(6):2786–2796

39. Boutouyrie P, Laurent S, Girerd X, Benetos A, Lacolley P, Abergel E, Safar M (1995) Common carotid artery stiffness and patterns of left ventricular hypertrophy in hypertensive patients. Hypertension 25(4 pt 1):651–659

40. Roman MJ, Saba PS, Pini R, Spitzer M, Pickering TG, Rosen S, Alderman MH, Devereux RB (1992) Parallel cardiac and vascular adaptation in hypertension. Circulation 86(6):1909–1918

41. Franklin SS, Gustin W, Wong ND, Larson MG, Weber MA, Kannel WB, Levy D (1997) Hemodynamic patterns of age-related changes in blood pressure. The Framingham Heart Study. Circulation 96(1):308–315

Chapter 2
Mechanisms of Arterial Stiffness

Abstract Current understanding for the mechanisms contributing to arterial stiffness is limited. The field is rapidly growing, however, and the complex process of functional, structural and signaling pathways working together to stiffen arteries is becoming increasingly clear. As such, arterial dilation and constriction, extracellular matrix accumulation and stiffening of individual cells via specific signal transduction pathways inter-connect providing numerous targets for potential therapeutic intervention. Herein we highlight the mechanisms that are largely implicated in arterial stiffness, and those that may be emerging as important targets.

Keywords Artery function • Smooth muscle • Adventitia • Endothelium • Collagen • Elastin • Advanced glycation end-products

Introduction

The mechanisms by which arteries stiffen include both physiological as well as cellular and molecular events that collectively contribute to the pathophysiology. Both functional and structural changes occur to stiffen arteries, which are due to the actions and interactions of the physiological and cellular/molecular events. In this chapter we will overview functional and structural changes, and highlight key cellular signaling mechanisms that promote these impairments. In order to gain greater insight for mechanisms contributing to arterial stiffness, we will use animal based studies to complement and to provide additional insight to support what is known from human-based literature. For example, aging mouse models are commonly used to study aortic stiffness, which demonstrate in vivo increases in aortic pulse wave velocity (aPWV) (Fig. 2.1a) and intrinsic stiffness assessed by ex vivo mechanical testing (Fig. 2.1b). Importantly, animal studies provide additional insight but the mechanisms do not always translate well to humans. A more in-depth understanding of mechanisms, however, ultimately leads to more efficacious treatments.

© Springer International Publishing Switzerland 2015

B.S. Fleenor, A.J. Berrones, *Arterial Stiffness*, SpringerBriefs in Physiology, DOI 10.1007/978-3-319-24844-8_2

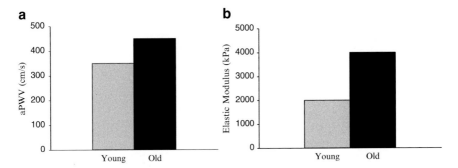

Fig. 2.1 A representative age-related increase of (**a**) in vivo arterial stiffness assessed by aortic pulse wave velocity (aPWV) and (**b**) ex vivo mechanical stiffness in isolated aortic segments of young (4–6 months) and old (26–28 months) mice

Functional Changes Contributing to Arterial Stiffness

Blood Pressure

Blood pressure is a traditional measure assessed clinically to determine risk for potential future cardiovascular events. Advancing age is associated with isolated systolic hypertension where current blood pressure control rates are suboptimal at ~50 % [1]. Aortic stiffness is considered to be a key mediator for controlling resistant hypertension. For instance, the Preterax in Regression of Arterial Stiffness (REASON) trial demonstrates that increased arterial stiffness is related to reduced blood pressure control, and decreases in arterial stiffness is a major contributor in the reduction and control of systolic blood pressure [2]. Additionally, other investigations have observed that increased aortic stiffness predicts incident hypertension and cardiovascular events [3, 4]. Collectively, current evidence indicates an effect for arterial stiffness to contribute to hypertension-related conditions.

In addition to the evidence for arterial stiffness to promote and regulate blood pressure, other data indicate blood pressure contributes to arterial stiffening. Recently, it has been proposed that "early vascular aging," which is, in part, characterized by aortic stiffness is present in young hypertensive adults [5]. The changes in arterial stiffness that are observed in young hypertensive subjects are similar to what is seen in older non-hypertensive adults. As such, when blood pressure is reduced in young adults with hypertension arterial stiffness is also decreased, suggesting blood pressure does indeed contribute to arterial stiffness early in life [6]. The current evidence indicates arterial stiffness and blood pressure promote and/or decrease the expression of one another in a feedback loop. Thus, blood pressure is a key mechanism by which aortic stiffness may be modulated to reduce cardiovascular risk.

Impaired Smooth Muscle Function

Vascular smooth muscle cell dysfunction results in impaired vasodilation, increased vasoconstriction, increased proliferation and migration, and has been postulated to contribute to the overall impairments in blood pressure regulation and arterial stiffness. In a recent meta-analysis, it was shown that smooth muscle dysfunction occurs with advancing age in adults [7]. A limitation of this analysis is that the effect of dysfunctional smooth muscle is relatively small, and was observed in peripheral arteries. Due to current limitations in technology, assessing smooth muscle function of the central arteries, such as the aorta, in adults cannot be performed readily. However, aortic segments from animals in age- and hypertension-related models have shown that there are indeed impairments in smooth muscle function [8, 9]. Thus, there is evidence to support the notion for impaired aortic smooth muscle function to be associated with age- and blood pressure-related vascular dysfunction, including arterial stiffness. Importantly, it has been estimated that ~50 % of aortic stiffness with aging is due to smooth muscle cells [10], which has been attributed to signaling and structural events in smooth muscle cells [10, 11]. These findings collectively indicate smooth muscle dysfunction, as it relates to extracellular signaling and structural changes, significantly contributes to the development of arterial stiffness.

Impaired Endothelium-Dependent Dilation

The vascular endothelium is a monolayer of cells on the innermost side of arteries that responds to mechanical forces and receptor-mediated signaling, contributing to arterial homeostasis and pathophysiology. Smooth muscle function is influenced by the vasoactive factors released from the vascular endothelium with nitric oxide (NO) being a primary signaling molecule. It is important to note, as discussed in the previous section, that endothelium-independent dilation (EID), or smooth muscle dysfunction, contributes much less to arterial dysfunction in comparison to endothelium dependent dilation (EDD) [7, 12]. Studies demonstrating greater smooth muscle dysfunction have largely been reported in subjects with additional cardiovascular risk factors [12]. This is further supported in animal models of aging where EID, or smooth muscle cell relaxation, is largely unimpaired in old mice when arterial segments are treated with the NO donor sodium nitroprusside [13–15]. Taken together, age-related impairments in EDD influences smooth muscle function—rather than influencing the dilatory properties—that is related to signaling and structural changes of the cells. Thus, targeting EDD via improvements in NO bioavailability may improve structural changes in vascular smooth muscle cells, which would be critical for reducing arterial stiffness (Fig. 2.2).

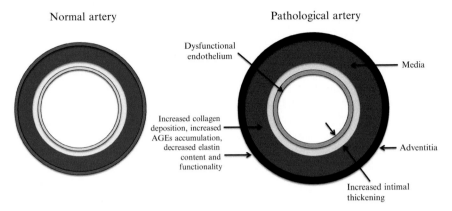

Fig. 2.2 A summary of the complex, multiple causes of arterial stiffness with aging and disease

Structural Changes Contributing to Arterial Stiffness

Aging results in multifaceted changes within the vasculature to promote arterial stiffness. These alterations include gross morphological remodeling and compositional changes within arteries that are not as readily identifiable in humans. Additionally, there are cell-specific changes that contribute to the overall decrements in arterial function and structure that have recently been elucidated. This section will highlight the importance of each structural component in the development of arterial stiffness (Fig. 2.2).

Gross Morphological Changes

Intima-media thickness (IMT) is a common clinical measure providing insight for an individual's vascular age and health. IMT increases as a product of aging, which is associated with increased stiffness and cardiovascular events [6]. IMT, a surrogate arterial stiffness measure is used clinically, but is more indicative of the gross changes in morphology rather than arterial stiffness. Moreover, IMT is not as sensitive of a predictor for major cardiovascular events in high-risk diabetic and/or hypertensive patients as other direct measures of arterial stiffness, [16] such as aPWV, because it only accounts for morphology, and excludes additional factors including material composition of the artery that is taken into account with aPWV [17]. The IMT thickening, however, is a relatively easy endpoint to assess for a trained clinical technician that provides additional insight for arterial health and future cardiovascular disease risk.

Compositional Changes

Collagen

Collagen is an important extracellular protein in arteries influencing arterial stiffness with over 20 identified isoforms to date [18]. The type I and III collagen isoforms are the most abundant in the aorta with type I having the greatest content throughout this large elastic artery [19]. This distribution pattern is important as type I collagen promotes strength leading to increased mechanical stiffness, whereas type III has greater elastic properties promoting elasticity. Type I collagen expression has an age-related increase in the three primary cell types of arteries: endothelial cells, vascular smooth muscle cells, and adventitial fibroblasts [20]. The compositional change for increased collagen I content and/or expression in each layer indicates the importance of this protein in arterial stiffening. For example, a collagenase-resistant animal model with the inability to decrease type I collagen was shown to have greater mechanical stiffness [21]. And, selective increases in adventitial collagen deposition, as observed with aging, may have even greater contributions to overall arterial stiffness compared to other arterial layers [22].

Type II collagen expression in arteries is an emerging isoform that may have important implications in age-related aortic stiffness. Arterial collagen II expression increases with aging in both rodents and humans [23]. This type II isoform is a matrix protein primarily found in cartilage, a biomaterial designed for strength. Thus, greater collagen II, in addition to increased collagen I protein expression in arteries is hypothesized to dramatically promote age-related arterial stiffening.

Elastin

The elasticity of arteries is primarily due to the extracellular protein elastin as it accounts for ~90 % of elastic fibers in arteries [18]. Aging and disease decrease elastin content and/or functionality leading to arterial stiffening as observed in the elastin haploinsufficient mouse model [24]. Elastin is expressed in all arterial cell types, but the medial smooth muscle cell layer expresses greater amounts of this protein [20]. Notably, arterial elastin expression progressively decreases with aging and is highest in early life. As elastin decreases, it is thought to be an irreparable process without the ability to re-form elastin in arteries. Prevention of elastin loss and functionality in early age would be of great importance in reducing arterial stiffness across the life span.

Advanced Glycation End-Products (AGEs)

Glycosylation, a post-translational modification of proteins, is also implicated in age- and disease-induced aortic stiffening. Non-specific accumulation and cross-linking of extracellular proteins by AGEs has been shown to contribute to arterial stiffness in both aged rodents and adults [20, 25]. The greater cross-linking by AGEs results in increased stiffness, which is akin to a chain link fence with more links having less flexibility. AGEs accumulation is hypothesized to be the result of one of three primary mechanisms, which include: (1) increased circulating glucose concentrations, (2) greater oxidative stress within arteries, and (3) enhanced inflammatory signaling [26]. Although these factors are all considered primary pathways by which arterial AGEs are formed, greater mechanistic insight for each pathway requires further investigation. In addition to the cross-linking effect of AGEs to promote arterial stiffness, the cellular signaling of AGEs is also a potential mechanism for arterial stiffening that will be discussed later in this chapter.

Calcium

Rigid deposits of calcium mineral within the vascular wall increase in vivo stiffness [27]. Coronary calcium levels are associated with those of the aorta and both are shown to increase with advancing age [28]. In a middle-aged cohort without prevalent cardiovascular disease from the Framingham Heart Study, a correlation was found between aortic stiffness and vascular calcification [29]. The appearance of progressive aortic stiffness and aortic calcification appears as early as the fourth decade of life [30]. Thus, calcification is an important mechanism leading to aortic stiffness that warrants further investigation.

Cellular Changes

There are distinct changes in vascular smooth muscle cell (VSMC) and endothelial cell stiffness that contribute to the overall arterial stiffening process [11, 31]. Increased VSMC stiffness has been attributed to changes in the cytoskeletal protein actin, as reductions in this protein decreases age-related VSMC stiffness [11]. Intracellular structural changes, in part, begin to explain how nearly 50 % of arterial stiffness is due to VSMCs. For instance, focal adhesions link the contractile components of the cell, including actin, to the extracellular matrix, which have been shown to mechanistically promote arterial stiffness [10]. Thus, the actin cytoskeleton and focal adhesions seem to collectively coordinate the VSMC-contributions to overall arterial stiffening.

Endothelial cells have also been shown to increase cellular stiffness in an animal model of obesity [31]. In this model, endothelial cell stiffness increased ~5 fold, similar to what is observed in VSMCs. Few studies have examined endothelial cell stiffness, and, therefore, the mechanistic insight by which these cells become stiffer is limited. Notwithstanding, a senescent endothelial cell model has indicated that smooth muscle alpha actin and collagen I protein expressions increase with aging, which would implicate endothelial cells as a contributor to arterial stiffness. Importantly, the pro-inflammatory cytokine tumor necrosis factor alpha was shown to recapitulate both smooth muscle alpha actin and collagen I protein expressions in non-senescent endothelial cells [32]. Hence, endothelial cell stiffness is an emerging contributor to arterial stiffness, which may be mediated by inflammatory processes.

In summary, cellular changes to both VSMCs and endothelial cells result in greater cell stiffness and provide novel insight for arterial stiffness. This novel area of investigation requires future study to determine the mechanisms by which VSMCs and endothelial cells stiffen with aging and disease.

Signaling Mechanisms

There are numerous potential signaling events that may contribute to age- and disease-related arterial stiffness. Our aim, however, is to highlight several key signaling and molecular mechanisms that have been shown to promote arterial stiffness. Thus, we present several important mechanisms shown to have a significant influence on arterial stiffness.

Oxidative Stress

Oxidative stress, which can be defined as increased reactive oxygen species (ROS) production or bioavailability in relation to the antioxidant buffering capacity, is greater in arteries of old animals. Increased NADPH oxidase (NOX) expression and/or activity is one oxidase system implicated in arterial ROS production with age, which is associated with decreased expression/activity of the superoxide dismutase (SOD) antioxidant defense system. As such, greater superoxide production within the aorta is associated with increased aortic stiffness observed with aging, and reductions in aortic superoxide bioavailability with antioxidant supplementation reverses arterial stiffening [13]. Importantly, antioxidant treatment is associated with an attenuation of age-related increases in the pro-oxidant NOX p67 subunit and reductions in the antioxidant manganese superoxide dismutase expressions [13–15, 33]. The oxidative stress process, however, is much more complex but currently there is little mechanistic evidence or insight. In short, greater oxidative stress within arteries promotes aortic stiffness via increased pro-oxidant superoxide production and reduced superoxide dismutation by the antioxidant system.

An important functional component of arterial stiffness is reduced nitric oxide (NO) bioavailability due to increased arterial ROS production with advancing age. Notably, antioxidant treatment reverses the age-related decrements in NO bioavailability, and is associated with reversal and/or attenuation with the structural and extracellular matrix components of arterial stiffening. For example, short-term (4-week treatment protocols) with SOD-specific boosting intervention, and less specific antioxidant interventions have been shown to modulate both collagen and AGEs [13–15]. Yet, few interventions have shown attenuation of the age-related decrements in elastin content. Additionally, it is largely unknown if antioxidants influence vascular smooth muscle and/or endothelial cellular stiffness. It appears, however, antioxidant compounds are a potential treatment strategy to ameliorate arterial stiffness.

Inflammation

Large artery inflammation is emerging as an important aspect of arterial aging, and aortic stiffening. For instance, increased macrophage numbers, and pro-inflammatory cytokine expressions have been observed in aortas from older mice. Importantly, expression of the nuclear factor kappa B (NFκB) transcription factor is also greater in aortas with aging, which may mediate both monocyte recruitment and pro-inflammatory cytokine expression. Thus, examining the role of NFκB in arterial stiffness, and identifying novel interventions to modulate NFκB expression are promising [20, 34]. For example, inhibition of NFκB in older animals improves endothelium-dependent dilation via a NO-dependent mechanism. This finding indirectly suggests NFκB contributes to age-related aortic stiffness by reducing NO bioavailability and impairing endothelial function. As such, anti-inflammatory interventions may lead to reductions in aortic stiffness.

Nitric Oxide (NO)

Impaired endothelium-dependent dilation is a key functional outcome contributing to arterial stiffness. Decreased expression and/or activity of endothelial NO synthase (eNOS)—a critical enzyme—leads to reduced NO bioavailability and endothelial dysfunction [35]. The reduced capacity for the arteries to dilate indicates VSMCs are not relaxing, which is largely due to reductions in NO bioavailability. Equally important is the influence of NO to modulate arterial inflammatory proteins and transforming growth factor Beta 1, which the latter is implicated in structural changes within arteries [36]. Thus, NO boosting interventions are quite important as both functional and structural changes within arteries are influenced by this critical signaling mechanism.

Reductions in NO bioavailability have been attributed to oxidative stress. More specifically, greater superoxide production reacts with NO to form peroxynitrite, which modifies proteins and is a common marker of oxidative stress. Additionally, the key cofactor for NO production, tetrahydrobiopterin (BH_4) becomes oxidized contributing to uncoupling of eNOS resulting in less NO bioavailability and more superoxide production [35]. The greater oxidative stress is due to both enhanced pro-oxidant and reduced antioxidant enzyme expression and/or activity. Targeting oxidative stress to improve NO bioavailability would be critical for improving both functional and structural mechanisms leading to arterial stiffness.

Transforming Growth Factor Beta 1 (TGF-β1)

TGF-β1 is a profibrotic cytokine involved with adventitial remodeling, and thus implicated in aortic stiffness. Adventitial TGF-β1 expression is increased in aged rodents that, in turn induces a phenotypic change of adventitial fibroblasts into pro-secretory myofibroblasts. This phenotypic transition results in greater collagen I protein expression in myofibroblasts, which is associated with increased adventitial collagen I expression and aortic stiffness. The collagen secretory phenotype induced by TGF-β1 also promotes superoxide production, which was shown to mediate the effects of TGF-β1 in fibroblasts [37]. Thus, TGF-β1 is an important profibrotic cytokine with increased age-related expression with limited in vivo evidence for its effects on arterial stiffness. Interventions focused on attenuating adventitial TGF-β1 signaling may be of importance as the adventitia is a key load-bearing layer of arteries.

Advanced Glycation End-Products (AGEs)

As discussed previously, AGEs are responsible for cross-linking extracellular proteins that ultimately contribute to aortic stiffness. Notably, AGEs also bind the receptor of advanced glycation end-products (RAGE) which results in intracellular signaling that promotes negative biological consequences such as greater arterial stiffness seen with aging [26]. Limited experimental evidence exists for this hypothesis; however, it has been shown that ex vivo administration of biologically active AGEs to arteries from young rodents enhances aortic stiffness [38]. AGEs accumulation within the artery was not assessed, but these findings indicate AGEs signaling promotes increases in mechanical stiffness. Moreover, an AGEs enriched diet increases pro-inflammatory cytokine expressions in adipocytes, which may be the downstream signaling event leading to greater mechanical stiffness [39]. However, little is currently known about AGEs signaling and arterial stiffness.

Summary

The mechanisms underlying arterial stiffness are multifaceted, complex and interrelated. Although mechanistic insight for arterial stiffness is accumulating, the scientific community is in the beginning stages of understanding the disease's etiology. Novel therapeutics to effectively treat and manage arterial stiffness will target the functional and structural components of this emerging risk factor.

References

 1. Go AS, Mozaffarian D, Roger VL, Benjamin EJ, Berry JD, Blaha MJ, Dai S, Ford ES, Fox CS, Franco S, Fullerton HJ, Gillespie C, Hailpern SM, Heit JA, Howard VJ, Huffman MD, Judd SE, Kissela BM, Kittner SJ, Lackland DT, Lichtman JH, Lisabeth LD, Mackey RH, Magid DJ, Marcus GM, Marelli A, Matchar DB, McGuire DK, Mohler ER III, Moy CS, Mussolino ME, Neumar RW, Nichol G, Pandey DK, Paynter NP, Reeves MJ, Sorlie PD, Stein J, Towfighi A, Turan TN, Virani SS, Wong ND, Woo D, Turner MB (2014) Executive summary: heart disease and stroke statistics—2014 update: a report from the American Heart Association. Circulation 129(3):399–410. doi:10.1161/01.cir.0000442015.53336.12
 2. Protogerou A, Blacher J, Stergiou GS, Achimastos A, Safar ME (2009) Blood pressure response under chronic antihypertensive drug therapy: the role of aortic stiffness in the REASON (Preterax in Regression of Arterial Stiffness in a Controlled Double-Blind) study. J Am Coll Cardiol 53(5):445–451. doi:10.1016/j.jacc.2008.09.046
 3. Mitchell GF, Hwang SJ, Vasan RS, Larson MG, Pencina MJ, Hamburg NM, Vita JA, Levy D, Benjamin EJ (2010) Arterial stiffness and cardiovascular events: the Framingham Heart Study. Circulation 121(4):505–511. doi:10.1161/circulationaha.109.886655
 4. Najjar SS, Scuteri A, Shetty V, Wright JG, Muller DC, Fleg JL, Spurgeon HP, Ferrucci L, Lakatta EG (2008) Pulse wave velocity is an independent predictor of the longitudinal increase in systolic blood pressure and of incident hypertension in the Baltimore Longitudinal Study of Aging. J Am Coll Cardiol 51(14):1377–1383. doi:10.1016/j.jacc.2007.10.065
 5. Kotsis V, Stabouli S, Karafillis I, Nilsson P (2011) Early vascular aging and the role of central blood pressure. J Hypertens 29(10):1847–1853. doi:10.1097/HJH.0b013e32834a4d9f
 6. Harvey A, Montezano AC, Touyz RM (2015) Vascular biology of ageing-implications in hypertension. J Mol Cell Cardiol 83:112–121. doi:10.1016/j.yjmcc.2015.04.011
 7. Montero D, Pierce GL, Stehouwer CD, Padilla J, Thijssen DH (2015) The impact of age on vascular smooth muscle function in humans. J Hypertens 33(3):445–453; discussion 453. doi:10.1097/hjh.0000000000000446
 8. Kloss S, Bouloumie A, Mulsch A (2000) Aging and chronic hypertension decrease expression of rat aortic soluble guanylyl cyclase. Hypertension 35(1 pt 1):43–47
 9. Moritoki H, Tanioka A, Maeshiba Y, Iwamoto T, Ishida Y, Araki H (1988) Age-associated decrease in histamine-induced vasodilation may be due to reduction of cyclic GMP formation. Br J Pharmacol 95(4):1015–1022
10. Gao YZ, Saphirstein RJ, Yamin R, Suki B, Morgan KG (2014) Aging impairs smooth muscle-mediated regulation of aortic stiffness: a defect in shock absorption function? Am J Physiol Heart Circ Physiol 307(8):H1252–H1261. doi:10.1152/ajpheart.00392.2014
11. Qiu H, Zhu Y, Sun Z, Trzeciakowski JP, Gansner M, Depre C, Resuello RR, Natividad FF, Hunter WC, Genin GM, Elson EL, Vatner DE, Meininger GA, Vatner SF (2010) Short communication: vascular smooth muscle cell stiffness as a mechanism for increased aortic stiffness with aging. Circ Res 107(5):615–619. doi:10.1161/circresaha.110.221846
12. Seals DR, Jablonski KL, Donato AJ (2011) Aging and vascular endothelial function in humans. Clin Sci (Lond) 120(9):357–375. doi:10.1042/cs20100476
13. Fleenor BS, Seals DR, Zigler ML, Sindler AL (2012) Superoxide-lowering therapy with TEMPOL reverses arterial dysfunction with aging in mice. Aging Cell 11(2):269–276. doi:10.1111/j.1474-9726.2011.00783.x
14. Fleenor BS, Sindler AL, Marvi NK, Howell KL, Zigler ML, Yoshizawa M, Seals DR (2013) Curcumin ameliorates arterial dysfunction and oxidative stress with aging. Exp Gerontol 48(2):269–276. doi:10.1016/j.exger.2012.10.008
15. Sindler AL, Fleenor BS, Calvert JW, Marshall KD, Zigler ML, Lefer DJ, Seals DR (2011) Nitrite supplementation reverses vascular endothelial dysfunction and large elastic artery stiffness with aging. Aging Cell 10(3):429–437. doi:10.1111/j.1474-9726.2011.00679.x
16. Laurent S, Girerd X, Mourad JJ, Lacolley P, Beck L, Boutouyrie P, Mignot JP, Safar M (1994) Elastic modulus of the radial artery wall material is not increased in patients with essential hypertension. Arterioscler Thromb 14(7):1223–1231

17. Saphirstein RJ, Morgan KG (2014) The contribution of vascular smooth muscle to aortic stiffness across length scales. Microcirculation 21(3):201–207. doi:10.1111/micc.12101
18. Diez J (2007) Arterial stiffness and extracellular matrix. Adv Cardiol 44:76–95. doi:10.1159/000096722
19. Tsamis A, Krawiec JT, Vorp DA (2013) Elastin and collagen fibre microstructure of the human aorta in ageing and disease: a review. J R Soc Interface 10(83):20121004. doi:10.1098/rsif.2012.1004
20. Fleenor BS (2013) Large elastic artery stiffness with aging: novel translational mechanisms and interventions. Aging Dis 4(2):76–83
21. Vafaie F, Yin H, O'Neil C, Nong Z, Watson A, Arpino JM, Chu MW, Wayne Holdsworth D, Gros R, Pickering JG (2014) Collagenase-resistant collagen promotes mouse aging and vascular cell senescence. Aging Cell 13(1):121–130. doi:10.1111/acel.12155
22. Schulze-Bauer CA, Regitnig P, Holzapfel GA (2002) Mechanics of the human femoral adventitia including the high-pressure response. Am J Physiol Heart Circ Physiol 282(6):H2427–H2440. doi:10.1152/ajpheart.00397.2001
23. Jiang L, Zhang J, Monticone RE, Telljohann R, Wu J, Wang M, Lakatta EG (2012) Calpain-1 regulation of matrix metalloproteinase 2 activity in vascular smooth muscle cells facilitates age-associated aortic wall calcification and fibrosis. Hypertension 60(5):1192–1199. doi:10.1161/hypertensionaha.112.196840
24. Wagenseil JE, Nerurkar NL, Knutsen RH, Okamoto RJ, Li DY, Mecham RP (2005) Effects of elastin haploinsufficiency on the mechanical behavior of mouse arteries. Am J Physiol Heart Circ Physiol 289(3):H1209–H1217. doi:10.1152/ajpheart.00046.2005
25. Semba RD, Najjar SS, Sun K, Lakatta EG, Ferrucci L (2009) Serum carboxymethyl-lysine, an advanced glycation end product, is associated with increased aortic pulse wave velocity in adults. Am J Hypertens 22(1):74–79. doi:10.1038/ajh.2008.320
26. Semba RD, Nicklett EJ, Ferrucci L (2010) Does accumulation of advanced glycation end products contribute to the aging phenotype? J Gerontol A Biol Sci Med Sci 65(9):963–975. doi:10.1093/gerona/glq074
27. Demer LL, Tintut Y (2014) Inflammatory, metabolic, and genetic mechanisms of vascular calcification. Arterioscler Thromb Vasc Biol 34(4):715–723. doi:10.1161/atvbaha.113.302070
28. Collins JA, Munoz JV, Patel TR, Loukas M, Tubbs RS (2014) The anatomy of the aging aorta. Clin Anat 27(3):463–466. doi:10.1002/ca.22384
29. Tsao CW, Pencina KM, Massaro JM, Benjamin EJ, Levy D, Vasan RS, Hoffmann U, O'Donnell CJ, Mitchell GF (2014) Cross-sectional relations of arterial stiffness, pressure pulsatility, wave reflection, and arterial calcification. Arterioscler Thromb Vasc Biol 34(11):2495–2500. doi:10.1161/atvbaha.114.303916
30. Sekikawa A, Shin C, Curb JD, Barinas-Mitchell E, Masaki K, El-Saed A, Seto TB, Mackey RH, Choo J, Fujiyoshi A, Miura K, Edmundowicz D, Kuller LH, Ueshima H, Sutton-Tyrrell K (2012) Aortic stiffness and calcification in men in a population-based international study. Atherosclerosis 222(2):473–477. doi:10.1016/j.atherosclerosis.2012.03.027
31. DeMarco VG, Habibi J, Jia G, Aroor AR, Ramirez-Perez FI, Martinez-Lemus LA, Bender SB, Garro M, Hayden MR, Sun Z, Meininger GA, Manrique C, Whaley-Connell A, Sowers JR (2015) Low-dose mineralocorticoid receptor blockade prevents western diet-induced arterial stiffening in female mice. Hypertension 66(1):99–107. doi:10.1161/hypertensionaha.115.05674
32. Fleenor BS, Marshall KD, Rippe C, Seals DR (2012) Replicative aging induces endothelial to mesenchymal transition in human aortic endothelial cells: potential role of inflammation. J Vasc Res 49(1):59–64. doi:10.1159/000329681
33. Hamilton CA, Brosnan MJ, McIntyre M, Graham D, Dominiczak AF (2001) Superoxide excess in hypertension and aging: a common cause of endothelial dysfunction. Hypertension 37(2 pt 2):529–534
34. Tan Y, Tseng PO, Wang D, Zhang H, Hunter K, Hertzberg J, Stenmark KR, Tan W (2014) Stiffening-induced high pulsatility flow activates endothelial inflammation via a TLR2/NF-kappaB pathway. PLoS One 9(7), e102195. doi:10.1371/journal.pone.0102195

35. Sindler AL, Reyes R, Chen B, Ghosh P, Gurovich AN, Kang LS, Cardounel AJ, Delp MD, Muller-Delp JM (2013) Age and exercise training alter signaling through reactive oxygen species in the endothelium of skeletal muscle arterioles. J Appl Physiol 114(5):681–693. doi:10.1152/japplphysiol.00341.2012
36. Koyanagi M, Egashira K, Kubo-Inoue M, Usui M, Kitamoto S, Tomita H, Shimokawa H, Takeshita A (2000) Role of transforming growth factor-beta1 in cardiovascular inflammatory changes induced by chronic inhibition of nitric oxide synthesis. Hypertension 35(1 pt 1):86–90
37. Fleenor BS, Marshall KD, Durrant JR, Lesniewski LA, Seals DR (2010) Arterial stiffening with ageing is associated with transforming growth factor-beta1-related changes in adventitial collagen: reversal by aerobic exercise. J Physiol 588(pt 20):3971–3982. doi:10.1113/jphysiol.2010.194753
38. Fleenor BS, Sindler AL, Eng JS, Nair DP, Dodson RB, Seals DR (2012) Sodium nitrite de-stiffening of large elastic arteries with aging: role of normalization of advanced glycation end-products. Exp Gerontol 47(8):588–594. doi:10.1016/j.exger.2012.05.004
39. Cai W, Ramdas M, Zhu L, Chen X, Striker GE, Vlassara H (2012) Oral advanced glycation endproducts (AGEs) promote insulin resistance and diabetes by depleting the antioxidant defenses AGE receptor-1 and sirtuin 1. Proc Natl Acad Sci U S A 109(39):15888–15893. doi:10.1073/pnas.1205847109

Chapter 3
Implications of Arterial Stiffness

Abstract The implications for increased arterial stiffness are just now beginning to be understood. However, there is still much unknown in this emerging field of study. In this chapter we overview several areas of clinical importance that influence arterial stiffness, which include aging, sex, race, body composition, and cardiorespiratory fitness, and the implications of these on cardiovascular health and target organ damage. These discussions merely highlight the complex interactions that occur with aging alone, which become increasingly convoluted with multiple pathologies. Thus, our discussion only begins to elucidate the very complex processes contributing to aortic stiffness and the overall health implications.

Keywords Heart • Kidney • Brain • VO_2 peak • Epidemiology • Aorta

Arterial Stiffness as a Cardiovascular Disease (CVD) Biomarker

CVD is the leading cause of mortality in the United States and represents an important target for intervention [1]. Although aortic pulse wave velocity (aPWV)—the gold standard measure of arterial stiffness—is a nascent clinical outcome measure, its prognostic value as an independent CVD biomarker to identify future CV events and all-cause mortalities is still emerging and being established [2–13]. Ultimately, the overall expectation of a CVD biomarker is to enhance the ability of the clinician to optimally manage the patient and reduce the risk of future CV events [14]. aPWV is in line with this expectation, and may therefore be used in conjunction with, and independent of traditional CVD risk factors to identify future CV events and all-cause mortalities in numerous pathological conditions. In the following section, we will examine the epidemiology of arterial stiffness in the context of aging, sex, race, body composition, and habitual exercise (Fig. 3.1). We will conclude with the implications for increased arterial stiffness in target organs such as the heart, the brain, and the kidneys.

© Springer International Publishing Switzerland 2015
B.S. Fleenor, A.J. Berrones, *Arterial Stiffness*, SpringerBriefs in Physiology,
DOI 10.1007/978-3-319-24844-8_3

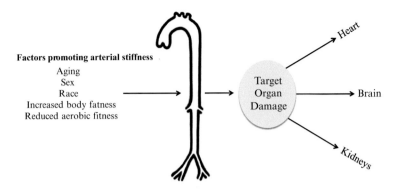

Fig. 3.1 Epidemiological contributors to arterial stiffness with potential implications for target organ damage

Epidemiology

Aging

In the United States, the proportion of patients greater than 65 years of age is increasing at a greater rate than the total population [15]. Importantly, the primary determinant of cardiovascular health is age [16]. Therefore understanding the impact of early, accelerated and/or heightened vascular aging are important initial steps for identifying vulnerable older populations. For instance, during 'normal' arterial aging there are functional, structural, and cellular and molecular events, which collectively results in arterial stiffness. However, individuals where the arterial aging process is accelerated beyond what would be expected in a normal or reference population, additional increases in aPWV will be observed indicating greater CVD risk [17]. Identifying the physiological mechanisms promoting age-related arterial stiffness, and the accelerating factors contributing to greater CVD risk are thus clinically significant.

The Baltimore Longitudinal Study on Aging (BLSA) is a large prospective cohort study that assessed aPWV over a period of 7 years in a subset of 943 participants [18]. Both older age and increased systolic blood pressure (SBP), or hypertension in this population, were the main longitudinal determinants of age-related increases in aPWV. Furthermore, the effect of SBP on the rate of increase in aPWV was evident even in prehypertensive subjects, and was associated with an accelerated rate of aPWV in a dose dependent manner. The isolated impact, however, of elevated SBP on arterial stiffness is difficult to partition and remains a current challenge. Notwithstanding, it has been shown that after controlling for traditional cardiovascular risk factors, including blood pressure, an increase in 5 m/s of aPWV is equivalent to aging 10 years, in terms of increased likelihood of CVD mortality [19]. In very old populations (i.e. 70–100 years), aPWV is a strong independent predictor of cardiovascular mortality with an adjusted odds ratio of 4.60 (95 % CI,

1.4–15.7) when aPWV was >17.7 m/s [20]. Thus, it is evident that both aging and even modest increases of SBP in older adults are primary factors in greater CVD risk in older populations.

To further highlight the influence of age on aortic stiffness in a large healthy normotensive population (age range = 18–90 years), it was shown that aPWV beyond 50 years increases at an accelerated rate [21]. Therefore, suggesting that arterial stiffness is a more sensitive marker of vascular aging in older compared with younger adults. Hence, in this study, the relationship between arterial stiffness showed an exponential increase after the fifth decade of life, and the inclusion of age as a predictor of aPWV accounted for 60 % of the total variance. Similarly in another study, after controlling for gender, body size, and heart rate, aPWV increased by 1.56 m/s per 10 years from age 20 to 80 years in which a marked trend occurred after 50 years of age [22].

Notably, in the presence of obesity and diabetes, arterial stiffness increases more rapidly with advancing age, which would be equivalent to a 70 year old adult [23, 24]. These findings provide additional insight for age as the primary factor influencing arterial stiffness, which is accelerated at ~50 years of age, and for both obesity and diabetes as contributors to accelerated age-induced aortic stiffness. Further, the National Institute on Aging SardiNIA study, a population-based, follow-up study involving more than 4000 community-living men and women (age range = 20–100 years) showed that aPWV increased by ~60 % from age 30 to 70 years [25]. An additional cross-sectional study using 102 Korean adults (age range = 21–60s years) demonstrated that aging accounted for 37 % of the total variance of arterial stiffness, which increased by 0.07 m/s per year [26]. In summary, arterial stiffness increases linearly until approximately 50 years of age, after which time there is an acceleration of the stiffening process.

Sex

Due to sex related differences in CVD incidence, prevalence, morbidity and mortality [27], it is important to discuss the influence of sex on the progression of arterial stiffness. For example, discrepancies exist between sexes regarding atherosclerosis, hypertension, angina, and stroke [27]. Therefore, predicting CVD risk in women is challenging because many statistical models of cardiovascular complications relying on traditional risk factors such as smoking, diabetes, and total cholesterol, have not accounted for sex differences that are known to exist [28].

Older women compared with aged men have a greater incidence of heart failure with preserved ejection fraction (HFpEF). Interestingly, aortic characteristic impedance (i.e. aortic opposition to pulsatile inflow from the contracting left ventricle, measured as the ratio of aortic pulsatile pressure to flow) has been shown to be greater in older women than older men, which is indicative of greater proximal ascending aortic stiffness and diastolic heart dysfunction in women [29]. But, aPWV is greater in men than women (11.9±3.8 vs. 10.5±3.4 m/s, P<0.0001) despite the fact that women had greater AIx% (surrogate measure of arterial stiffness) and higher central (aortic) SBP. These data suggest women may be more susceptible for presenting with a greater pulsatile load on the left ventricle resulting in diastolic dysfunction and an abnormal ventricular-arterial interaction leading to the increased incidence of HFpEF [29]. In this view, the role of pulse pressure amplification could be considered a potential mechanism to explain why women have greater aortic characteristic imped-ance, or proximal aortic stiffness, compared to men that is independent of aPWV, which largely does not account for stiffening of the ascending aorta.

In support of greater pulse pressure amplification as a sex-related mechanism, differences in the brachial pulse pressure to central pulse pressure ratio (B-PP/C-PP) were shown in older, post-menopausal women ($\geq$55 years) compared with older men [30]. More specifically, older women have a decreased B-PP to C-PP ratio because of greater increases in aortic pulse pressure than what is observed in men. Increased aortic stiffness, likely due to the stiffening of proximal ascending aorta, is responsible for the reduction in the B-PP to C-PP ratio in women but not men, who appear to maintain pulse pressure amplification with older age.

In fact, CVD risk is increased in older, post-menopausal women due to increased pulse pressure at the central (aortic) level [30]. Similarly, the SardiNIA showed that B-PP increased for both men and women at the entry age of 55 years, however for men the increase was attenuated until 70 years, which there was little change in B-PP after this age [25]. In contrast to men, the average longitudinal rate of change of B-PP for women continued to increase (monotonically) with every decade from age of 30 to beyond 70 years. In short, both B-PP and C-PP in women have shown to increase with advancing age compared with men [25, 30], indicating central blood pressure is a key target to reduce CVD risk in older women.

In addition to blood pressure, accelerated stiffening of the aortic and carotid arteries in older age (50+ years) has shown sex-related differences. The Multi-Ethnic Study of Atherosclerosis (MESA) project has demonstrated that women in the 45–54 age range had equivalent Young's elastic modulus (arterial stiffness) of the carotid artery compared to men of the same age [31]. However, in the oldest group (age range=75–84 years), women had significantly higher Young's elastic modulus compared to men (2013.7 vs. 1741.3 mmHg, $P < 0.05$). These data suggest that, following menopause, carotid artery stiffness progression increases more quickly in women than in men [31]. However, in the BLSA study, the longitudinal rate of change of aPWV was greater in men than in women, which may indicate that women display a differing pattern of arterial remodeling in the aortic and carotid regions compared with men [18]. Clearly, accelerated older age differences in arterial stiffness are present, which warrant further investigation in order clarify sex- and artery-specific differences on the progression of arterial stiffness.

Race

Predictions have suggested that over the next 35 years non-Hispanic Caucasians will no longer be the majority race in the United States, due to the increased number of Hispanics and Asians [32]. Given the changing demographics of the United States, understanding varied risk factor profiles and health outcomes will contribute to continued improvements of care for both minority and non-minority groups [32]. African Americans, for instance, experience a disproportionately greater burden of CVD, morbidity and mortality, relative to their Caucasian counterparts [1]. While this uneven burden of CVD between races has yet to be explained, emerging models point to social, environmental, and economic risks as accelerators of developmental processes, which become biological patterns that ultimately influence health and disease [33].

In the context of arterial stiffness, young (23 years) African Americans compared with Caucasians have greater aPWV (7.3 ± 0.3 vs. 6.0 ± 0.2 m/s; $P < 0.001$) that is independent of peripheral (brachial) blood pressure values, cardiorespiratory fitness, total cholesterol, fasting glucose, body fat, and glomerular filtration rate [34]. Additionally, African Americans have greater carotid intima-media thickness, aortic SBP, and reduced pulse pressure amplification. The increased arterial stiffness between African Americans and Caucasians suggests that large elastic artery stiffness and central blood pressure mechanisms are key contributors in CVD progression in the young African American population.

In middle-aged African Americans and Caucasians it was shown that African Americans had greater wave reflections and arterial stiffness in large elastic arteries [35]. More specifically, AIx (wave reflections) and aPWV were higher in African Americans compared with Caucasians. Similarly, in a middle-aged cohort of African American compared with Caucasian adults of the Atherosclerosis Risk in Communities Study (ARIC study) displayed earlier and more accelerated large artery stiffening (common carotids), which was associated with the development of hypertension [36]. Therefore, it seems apparent, racial disparities in arterial stiffness may help to explain why incident CVD occurs at an earlier age in African Americans compared with Caucasians [37]. Furthermore, treatments aimed at attenuating increased arterial stiffness where there is no apparent increase in blood pressure in young African Americans seems recommended in order to reduce accelerated CVD-related risk and adverse CV events in this population.

Body Composition

Assessment of body composition, particularly body fatness, and relating to measures of arterial stiffness is critical. However, there are many facets, and limitations for the different measures of body composition that need to be considered prior to discussing body composition and arterial stiffness. For example, fat and fat-free mass can be assessed by dual energy x-ray absorptiometry (DXA) and computed tomography (CT) scans that provide valid measures. Both DXA and CT scans are preferable measures, however, they are costly and are not always feasible for larger population studies. For larger studies, body mass index (BMI) is a common measure of total body fatness, which is determined with the ratio of weight (kg) to height (m^2). As such, BMI quantifies total mass, and does not selectively quantify fat or fat-free mass. Hence, individuals who have a high BMI may not necessarily have a high body fat percentage, but instead could have a large amount of fat-free mass that could increase BMI but not reflect greater fat mass. In large part, however, BMI is a standard and valid method for assessing body fatness in sedentary and recreational active individuals, or those without large amounts of muscle mass. Thus, understanding the limitations to BMI is important prior to reviewing studies reporting BMI but not other measures of body composition.

Greater fat mass resulting in obesity has been implicated in arterial stiffness across the lifespan. As such, overweight and obese adolescents (n = 86; BMI ≥ 85th percentile) demonstrate a higher aortic and brachial SBP, brachial and aortic PP, and mean arterial pressure (MAP), than adolescents of normal weight (n = 141; BMI < 85th percentile) [38]. Moreover, overweight/obese adolescents have a 7 % higher aPWV and 3.5 % lower PP amplification than the adolescents of the normal weight group [38]. It is particularly worrisome and likely that, given the negative implications of childhood obesity on arterial function, overweight and obese adolescents will carry unhealthy juvenile behaviors into adulthood that will give rise to early CVD risk. For instance, it has been shown in a large follow-up study that childhood body size or adiposity was associated with increased intima-media thickness and large artery stiffness as an adult [39]. Thus, it is imperative to develop interventions to attenuate adolescent arterial stiffening that is due to obesity.

Overeating, or excessive caloric intake, and/or a sedentary lifestyle contribute to weight gain and obesity. When normal weight, non-obese adults were overfed for 6–8 weeks by ~1000 kcal/day (5 kg of weight gain), total abdominal fat, abdominal visceral fat, and waist circumference increased in parallel with arterial stiffness by 13 % [40]. These findings were similar in Amsterdam Growth and Health Longitudinal Study, in which they found that an increase of trunk mass by 10 kg was positively associated with carotid Young's elastic modulus [41]. In short, the magnitude of abdominal visceral fat accumulation appears to be an important contributor of large artery stiffening in adults [40, 42].

In addition to short-term weight gain, chronic obesity in adults results in significant metabolic burden due to excessive fat storage, and ultimately, a clinical manifestation of arterial dysfunction. In large part, overweightness and obesity are considered to be important in arterial stiffening [40, 43, 44]. Contrary to this notion, it has been asserted that BMI is not very useful in predicting changes in arterial stiffness due to generalized obesity [45], and that the cardiovascular system of older adults may display an adverse association of body fat and arterial stiffness in contrast to young adults who may be more adaptable to the state of obesity [46]. Waist circumference, in fact, was the only measurement positively associated with early atherosclerosis and arterial stiffness in the Supplementation en Vitamines et Mineraux Antioxydants (SU.VI.MAX) study, in which over 1000 middle-aged adults were assessed for body composition and carotid structure and function [47]. In support of this finding, older adults from the Health ABC study demonstrated an association of aPWV and abdominal visceral fat (the strongest predictor of arterial stiffness) that was consistent across tertiles of body weight [48]. Additionally, older men with high amounts of total lean mass and low total fat mass exhibited the most favorable arterial profile, including reduced aPWV [49]. For older women, greater trunk fat mass was the strongest predictor of arterial stiffness [50]. Although general obesity does contribute to arterial stiffness for the population at large, some studies indicate abdominal obesity is a better predictor of arterial stiffness.

Habitual Exercise Training

Cardiorespiratory Fitness

Aging is the strongest predictor of arterial stiffness for both women and men. Chronic, regular physical activity can blunt the age-related arterial stiffness, as indicated in physically active post-menopausal women with similar aPWV compared to pre-menopausal women [51]. However, a significant difference in aPWV was still observed between post vs. pre-menopausal sedentary women indicating regular activity cannot fully reverse the age-related increase in arterial stiffness to those of younger adults. The strongest predictors of aPWV in this cohort were cardiorespiratory fitness (i.e. VO_2peak) and total and LDL cholesterol, which collectively explained up to 50 % of the variance in central arterial stiffness [51]. Similarly, VO_2peak has been shown to be associated with reduced PWV and body fatness, and greater amounts of regular physical activity in an apparently healthy middle aged cohort of women [52]. In addition, an important mechanistic relationship was shown in older but not younger women between VO_2peak and relative left ventricular wall thickness ($R = -0.32$, $P < 0.05$), indicating that more favorable cardiac remodeling is predicated on greater cardiorespiratory fitness. In brief, greater physical activity and improved cardiorespiratory fitness is associated with reduced arterial stiffness and more favorable cardiac remodeling in middle-aged and older women.

Men with higher cardiorespiratory fitness also demonstrate reduced arterial stiffness, in part, through reductions in resting heart rate [53]. In addition, it has been shown that men who engage in regular endurance exercise had lower aPWV than both resistance trained and sedentary (non-exercising) participants [54]. This relationship of reduced arterial stiffness was dependent on greater cardiorespiratory fitness, reduced SBP, and reduced endothelin-1, a vasoconstrictor, in the endurance trained men [54]. Importantly, it has also been demonstrated, for both men and women, greater cardiorespiratory fitness and reduced arterial stiffness is associated with improved occipitoparietal perfusion and better cognitive composite scores (memory and attention-executive function) than those with lower cardiorespiratory fitness and greater arterial stiffness [55]. Thus, suggesting cardiorespiratory fitness and reduced arterial stiffness may influence important cognition centers in the brain. Collectively, these data indicate chronic exercise and/or increased cardiorespiratory fitness in men and women are important for attenuating the aging associated effects of arterial stiffness.

Strength Training

Regular, progressive strength training is an important exercise modality for preserving the aging associated decrements in lean muscle mass, strength, and physical function. However, reports have suggested strength training may increase measures of arterial stiffness. For example, Bertovic et al. [56] compared apparently healthy young males who had regularly strength trained for 12 months to sedentary controls, and discovered that strength trained athletes had higher aortic characteristic impedance and β stiffness. Additionally, strength trained athletes had greater carotid and brachial pulse pressures compared with the sedentary controls [56]. Similarly, young men (age range = 20–38 years) who engaged in resistance training for 4 months increased β stiffness and decreased carotid arterial compliance was negatively related to changes in left ventricular mass and hypertrophy indices [57]. Interestingly, detraining reversed the negative impact of strength training on arterial compliance, as values returned to baseline after 4 months of detraining [57].

Last, it has been shown that young men (age = 21 years) who had been strength training for 2 years had greater aPWV, and increased plasma endothelin-1 levels, compared with endurance training athletes and sedentary controls [54]. Additionally, in middle aged men, strength trained vs. sedentary controls had greater β stiffness, SBP and MAP, while vasoreactivity to a sympathetic stimulus was similar between groups, indicating endothelial function remains intact [58]. In short, there is convincing evidence that strength training increases arterial stiffness in both young and middle aged men. However, it is important to note that strength training is a recommended form of exercise to prevent age-related decreases in muscle mass, and that concomitant aerobic exercise offsets the influence of strength training on arterial stiffness [59, 60]. Undeniably, strength training per current guidelines in conjunction with aerobic exercise is recommended.

Target Organ Damage

Increased arterial stiffness will increase the potential for damage in high-flow organs such as the heart, the brain, and the kidneys [61]. In brief, target (or end) organ damage in these high-flow organs is caused by greater loads of pulsatile energy that impair target organ microvascular function leading to ischemia, reduction in cognitive scores, and/or reduced glomerular filtration rate [61]. In this section we will briefly highlight emerging areas of investigation for arterial stiffness in target organ damage of the heart, brain and kidneys.

The Heart

In hypertensive patients, an increased media to lumen ratio is evident in small arteries (lumen diameter = 100–350 μM), followed by endothelial dysfunction and left ventricular hypertrophy (in fewer patients), independent of proteinuria [62]. Small artery remodeling has been implicated in the pathogenesis of hypertension, which may precede most clinical manifestations of target organ damage in mild essential hypertension [62]. In the Strong Heart Study, elevated fibrinogen levels was shown to be an independent correlate of cardiovascular target organ damage such as left ventricular hypertrophy and arterial stiffness [63]. Furthermore, stress-corrected midwall shortening, an estimate of myocardial contractility, was reduced in individuals in the highest tertile of fibrinogen, in addition to reductions in ejection fraction [63]. However, heart failure patients with preserved ejection fraction have increased central aortic stiffness, which explains the abnormal ventricular-vascular coupling beyond myocardial relaxation velocity and other measures of diastolic function [64]. Thus, the pathophysiological link between early cardiac abnormalities and arterial stiffness may be elevated systemic inflammation and central blood pressure [65, 66].

The Brain

In a large cohort of older adults (n = 812), of whom 71 % had hypertension, increased aPWV was shown to be independently associated with a greater burden of subclinical disease in cerebral arterial beds [67]. After adjustment for age, sex, conventional CVD risk factors, and pulse pressure, increased aPWV was positively linked to volume of white matter hyperintensity in the brain, suggesting a potential link for arterial stiffness to vascular dementia [67]. Similarly, greater aPWV was associated with increased risk for silent subcortical infarcts, white matter hyperintensity volume, and lower memory scores, indicating that progressive stiffening of the aorta reduced the wave reflection interface between the carotid and aorta permitted a greater load of pulsatile energy into the microvasculature of the brain [68]. Notably, a recent investigation in patients who underwent aortic valve replacement demonstrated that N-methyl-D-aspartate antibody (NR2Ab), a serum ischemic brain injury biomarker, was significantly higher in the aPWV-high group, thus highlighting a potential biomarker for cerebrovascular outcomes [69]. In contrast, however, others have shown arterial stiffness was not linked to white matter hyperintensity, but aortic arch PWV was positively linked to lacunar brain infarcts, after correction for confounding factors such as age, sex, hypertension duration [70]. Although there have been a few studies linking arterial stiffness to brain pathology, additional investigation is required to further elucidate this relation.

The Kidneys

Because the renal circulation provides relatively low resistance (impedance), greater pulsatile energy renders the glomerulus susceptible to pressure related damage [71]. Elevated aPWV is associated with reduced glomerular filtration rate (GFR), which may be mediated by elevated blood pressure [72]. This relationship of reduced GFR and elevated aPWV has been demonstrated in type 2 diabetics [72] and patients with chronic kidney disease [73]. Arterial stiffness, as measured by intima-media thickness is associated with albuminuria, another consequence of greater pulsatile energy within the renal vasculature [74]. For normotensive and hypertensive subjects, greater intima-media thickness was associated with greater renal resistive index values, indicating renal arterial disease (measured as [peak systolic velocity – end diastolic velocity]/[peak systolic velocity]). In renal and vascular tissues, accumulation of advanced glycation end products cause abnormal cross-linking, which, in conjunction with greater pressure related damage manifest a deterioration of kidney function during conditions of elevated arterial stiffness [75]. Collectively, arterial stiffness is implicated in kidney dysfunction but further study is needed.

Summary

Arterial stiffness is a CVD biomarker that can be used to improve patient disease risk stratification in conjunction with, or independent of traditional risk factors such as smoking, diabetes, and total cholesterol. Aging is a natural biological process that increases arterial stiffness beyond the third decade of life until mortality where known sex differences exist. Moreover, increased body fatness is associated with early and accelerated arterial stiffness adding to the complexity. Importantly, habitual exercise training resulting in improvements in cardiorespiratory fitness is associated with reductions in arterial stiffness; whereas, strength training appears to increase arterial stiffness, which can be offset with aerobic exercise. Collectively, factors increasing arterial stiffness are associated with target organ damage to tissues such as the heart, the brain, and the kidney. Factors leading to greater aortic stiffness need to be identified to reduce CVD-related risk, and damage to critical organs.

References

1. Mozaffarian D, Benjamin EJ, Go AS, Arnett DK, Blaha MJ, Cushman M, de Ferranti S, Despres JP, Fullerton HJ, Howard VJ, Huffman MD, Judd SE, Kissela BM, Lackland DT, Lichtman JH, Lisabeth LD, Liu S, Mackey RH, Matchar DB, McGuire DK, Mohler ER III, Moy CS, Muntner P, Mussolino ME, Nasir K, Neumar RW, Nichol G, Palaniappan L, Pandey DK, Reeves MJ, Rodriguez CJ, Sorlie PD, Stein J, Towfighi A, Turan TN, Virani SS, Willey JZ, Woo D, Yeh RW, Turner MB (2015) Heart disease and stroke statistics—2015 update: a report from the American Heart Association. Circulation 131(4):e29–e322. doi:10.1161/cir.0000000000000152
2. Ben-Shlomo Y, Spears M, Boustred C, May M, Anderson SG, Benjamin EJ, Boutouyrie P, Cameron J, Chen CH, Cruickshank JK, Hwang SJ, Lakatta EG, Laurent S, Maldonado J, Mitchell GF, Najjar SS, Newman AB, Ohishi M, Pannier B, Pereira T, Vasan RS, Shokawa T, Sutton-Tyrell K, Verbeke F, Wang KL, Webb DJ, Willum Hansen T, Zoungas S, McEniery CM, Cockcroft JR, Wilkinson IB (2014) Aortic pulse wave velocity improves cardiovascular event prediction: an individual participant meta-analysis of prospective observational data from 17,635 subjects. J Am Coll Cardiol 63(7):636–646. doi:10.1016/j.jacc.2013.09.063
3. Blacher J, Asmar R, Djane S, London GM, Safar ME (1999) Aortic pulse wave velocity as a marker of cardiovascular risk in hypertensive patients. Hypertension 33(5):1111–1117

4. Boutouyrie P, Tropeano AI, Asmar R, Gautier I, Benetos A, Lacolley P, Laurent S (2002) Aortic stiffness is an independent predictor of primary coronary events in hypertensive patients: a longitudinal study. Hypertension 39(1):10–15

5. Koivistoinen T, Virtanen M, Hutri-Kahonen N, Lehtimaki T, Jula A, Juonala M, Moilanen L, Aatola H, Hyttinen J, Viikari JS, Raitakari OT, Kahonen M (2012) Arterial pulse wave velocity in relation to carotid intima-media thickness, brachial flow-mediated dilation and carotid artery distensibility: the Cardiovascular Risk in Young Finns Study and the Health 2000 Survey. Atherosclerosis 220(2):387–393. doi:10.1016/j.atherosclerosis.2011.08.007

6. Laurent S, Katsahian S, Fassot C, Tropeano AI, Gautier I, Laloux B, Boutouyrie P (2003) Aortic stiffness is an independent predictor of fatal stroke in essential hypertension. Stroke 34(5):1203–1206. doi:10.1161/01.str.0000065428.03209.64

7. Mattace-Raso FU, van der Cammen TJ, Hofman A, van Popele NM, Bos ML, Schalekamp MA, Asmar R, Reneman RS, Hoeks AP, Breteler MM, Witteman JC (2006) Arterial stiffness and risk of coronary heart disease and stroke: the Rotterdam Study. Circulation 113(5):657–663. doi:10.1161/circulationaha.105.555235

8. Mitchell GF, Hwang SJ, Vasan RS, Larson MG, Pencina MJ, Hamburg NM, Vita JA, Levy D, Benjamin EJ (2010) Arterial stiffness and cardiovascular events: the Framingham Heart Study. Circulation 121(4):505–511. doi:10.1161/circulationaha.109.886655

9. Pereira T, Maldonado J, Pereira L, Conde J (2013) Aortic stiffness is an independent predictor of stroke in hypertensive patients. Arq Bras Cardiol 100(5):437–443. doi:10.5935/abc.20130079

10. Pereira T, Maldonado J, Polonia J, Silva JA, Morais J, Rodrigues T, Marques M (2014) Aortic pulse wave velocity and HeartSCORE: improving cardiovascular risk stratification. A sub-analysis of the EDIVA (Estudo de DIstensibilidade VAscular) project. Blood Press 23(2):109–115. doi:10.3109/08037051.2013.823760

11. Regnault V, Lagrange J, Pizard A, Safar ME, Fay R, Pitt B, Challande P, Rossignol P, Zannad F, Lacolley P (2014) Opposite predictive value of pulse pressure and aortic pulse wave velocity on heart failure with reduced left ventricular ejection fraction: insights from an Eplerenone Post-Acute Myocardial Infarction Heart Failure Efficacy and Survival Study (EPHESUS) sub-study. Hypertension 63(1):105–111. doi:10.1161/hypertensionaha.113.02046

12. Vlachopoulos C, Aznaouridis K, Stefanadis C (2010) Prediction of cardiovascular events and all-cause mortality with arterial stiffness: a systematic review and meta-analysis. J Am Coll Cardiol 55(13):1318–1327. doi:10.1016/j.jacc.2009.10.061

13. Vlachopoulos C, Ioakeimidis N, Aznaouridis K, Terentes-Printzios D, Rokkas K, Aggelis A, Panagiotakos D, Stefanadis C (2014) Prediction of cardiovascular events with aortic stiffness in patients with erectile dysfunction. Hypertension 64(3):672–678. doi:10.1161/hypertensionaha.114.03369

14. Vasan RS (2006) Biomarkers of cardiovascular disease: molecular basis and practical considerations. Circulation 113(19):2335–2362. doi:10.1161/circulationaha.104.482570

15. Vigen R, Maddox TM, Allen LA (2012) Aging of the United States population: impact on heart failure. Curr Heart Fail Rep 9(4):369–374. doi:10.1007/s11897-012-0114-8

16. North BJ, Sinclair DA (2012) The intersection between aging and cardiovascular disease. Circ Res 110(8):1097–1108. doi:10.1161/circresaha.111.246876

17. Laurent S (2012) Defining vascular aging and cardiovascular risk. J Hypertens 30(suppl):S3–S8. doi:10.1097/HJH.0b013e328353e501

18. AlGhatrif M, Strait JB, Morrell CH, Canepa M, Wright J, Elango P, Scuteri A, Najjar SS, Ferrucci L, Lakatta EG (2013) Longitudinal trajectories of arterial stiffness and the role of blood pressure: the Baltimore Longitudinal Study of Aging. Hypertension 62(5):934–941. doi:10.1161/hypertensionaha.113.01445

19. Laurent S, Boutouyrie P, Asmar R, Gautier I, Laloux B, Guize L, Ducimetiere P, Benetos A (2001) Aortic stiffness is an independent predictor of all-cause and cardiovascular mortality in hypertensive patients. Hypertension 37(5):1236–1241

20. Meaume S, Benetos A, Henry OF, Rudnichi A, Safar ME (2001) Aortic pulse wave velocity predicts cardiovascular mortality in subjects >70 years of age. Arterioscler Thromb Vasc Biol 21(12):2046–2050

21. McEniery CM, Yasmin, Hall IR, Qasem A, Wilkinson IB, Cockcroft JR (2005) Normal vascular aging: differential effects on wave reflection and aortic pulse wave velocity: the Anglo-Cardiff Collaborative Trial (ACCT). J Am Coll Cardiol 46(9):1753–1760. doi:10.1016/j.jacc.2005.07.037

22. Redheuil A, Yu WC, Wu CO, Mousseaux E, de Cesare A, Yan R, Kachenoura N, Bluemke D, Lima JA (2010) Reduced ascending aortic strain and distensibility: earliest manifestations of vascular aging in humans. Hypertension 55(2):319–326. doi:10.1161/hypertensionaha.109.141275

23. Mitchell GF, Guo CY, Benjamin EJ, Larson MG, Keyes MJ, Vita JA, Vasan RS, Levy D (2007) Cross-sectional correlates of increased aortic stiffness in the community: the Framingham Heart Study. Circulation 115(20):2628–2636. doi:10.1161/circulationaha.106.667733

24. Gepner AD, Korcarz CE, Colangelo LA, Hom EK, Tattersall MC, Astor BC, Kaufman JD, Liu K, Stein JH (2014) Longitudinal effects of a decade of aging on carotid artery stiffness: the multiethnic study of atherosclerosis. Stroke 45(1):48–53. doi:10.1161/strokeaha.113.002649

25. Scuteri A, Morrell CH, Orru M, Strait JB, Tarasov KV, Ferreli LA, Loi F, Pilia MG, Delitala A, Spurgeon H, Najjar SS, AlGhatrif M, Lakatta EG (2014) Longitudinal perspective on the conundrum of central arterial stiffness, blood pressure, and aging. Hypertension 64(6):1219–1227. doi:10.1161/hypertensionaha.114.04127

26. Logan JG, Barksdale DJ (2013) Pulse wave velocity in Korean American men and women. J Cardiovasc Nurs 28(1):90–96. doi:10.1097/JCN.0b013e3182376685

27. Winham SJ, de Andrade M, Miller VM (2015) Genetics of cardiovascular disease: importance of sex and ethnicity. Atherosclerosis. doi:10.1016/j.atherosclerosis.2015.03.021

28. Manson JE, Bassuk SS (2015) Biomarkers of cardiovascular disease risk in women. Metabolism 64(3 suppl 1):S33–S39. doi:10.1016/j.metabol.2014.10.028

29. Coutinho T, Borlaug BA, Pellikka PA, Turner ST, Kullo IJ (2013) Sex differences in arterial stiffness and ventricular-arterial interactions. J Am Coll Cardiol 61(1):96–103. doi:10.1016/j.jacc.2012.08.997

30. Regnault V, Thomas F, Safar ME, Osborne-Pellegrin M, Khalil RA, Pannier B, Lacolley P (2012) Sex difference in cardiovascular risk: role of pulse pressure amplification. J Am Coll Cardiol 59(20):1771–1777. doi:10.1016/j.jacc.2012.01.044

31. Stern R, Tattersall MC, Gepner AD, Korcarz CE, Kaufman J, Colangelo LA, Liu K, Stein JH (2015) Sex differences in predictors of longitudinal changes in carotid artery stiffness: the Multi-Ethnic Study of Atherosclerosis. Arterioscler Thromb Vasc Biol 35(2):478–484. doi:10.1161/atvbaha.114.304870

32. Graham G (2014) Population-based approaches to understanding disparities in cardiovascular disease risk in the United States. Int J Gen Med 7:393–400. doi:10.2147/ijgm.s65528

33. Kuzawa CW, Sweet E (2009) Epigenetics and the embodiment of race: developmental origins of US racial disparities in cardiovascular health. Am J Hum Biol 21(1):2–15. doi:10.1002/ajhb.20822

34. Heffernan KS, Jae SY, Wilund KR, Woods JA, Fernhall B (2008) Racial differences in central blood pressure and vascular function in young men. Am J Physiol Heart Circ Physiol 295(6):H2380–H2387. doi:10.1152/ajpheart.00902.2008

35. Morris AA, Patel RS, Binongo JN, Poole J, Al Mheid I, Ahmed Y, Stoyanova N, Vaccarino V, Din-Dzietham R, Gibbons GH, Quyyumi A (2013) Racial differences in arterial stiffness and microcirculatory function between Black and White Americans. J Am Heart Assoc 2(2), e002154. doi:10.1161/jaha.112.002154

36. Din-Dzietham R, Couper D, Evans G, Arnett DK, Jones DW (2004) Arterial stiffness is greater in African Americans than in whites: evidence from the Forsyth County, North Carolina, ARIC cohort. Am J Hypertens 17(4):304–313. doi:10.1016/j.amjhyper.2003.12.004

37. Feinstein M, Ning H, Kang J, Bertoni A, Carnethon M, Lloyd-Jones DM (2012) Racial differences in risks for first cardiovascular events and noncardiovascular death: the Atherosclerosis Risk in Communities study, the Cardiovascular Health Study, and the Multi-Ethnic Study of Atherosclerosis. Circulation 126(1):50–59. doi:10.1161/circulationaha.111.057232

38. Pierce GL, Zhu H, Darracott K, Edet I, Bhagatwala J, Huang Y, Dong Y (2013) Arterial stiffness and pulse-pressure amplification in overweight/obese African-American adolescents: relation with higher systolic and pulse pressure. Am J Hypertens 26(1):20–26. doi:10.1093/ajh/hps014

39. Huynh Q, Blizzard L, Sharman J, Magnussen C, Schmidt M, Dwyer T, Venn A (2013) Relative contributions of adiposity in childhood and adulthood to vascular health of young adults. Atherosclerosis 228(1):259–264. doi:10.1016/j.atherosclerosis.2013.02.022

40. Orr JS, Gentile CL, Davy BM, Davy KP (2008) Large artery stiffening with weight gain in humans: role of visceral fat accumulation. Hypertension 51(6):1519–1524. doi:10.1161/hypertensionaha.108.112946

41. Schouten F, Twisk JW, de Boer MR, Stehouwer CD, Serne EH, Smulders YM, Ferreira I (2011) Increases in central fat mass and decreases in peripheral fat mass are associated with accelerated arterial stiffening in healthy adults: the Amsterdam Growth and Health Longitudinal Study. Am J Clin Nutr 94(1):40–48. doi:10.3945/ajcn.111.013532

42. Strasser B, Arvandi M, Pasha EP, Haley AP, Stanforth P, Tanaka H (2015) Abdominal obesity is associated with arterial stiffness in middle-aged adults. Nutr Metab Cardiovasc Dis 25(5):495–502. doi:10.1016/j.numecd.2015.01.002

43. Anoop S, Misra A, Bhardwaj S, Gulati S (2015) High body fat and low muscle mass are associated with increased arterial stiffness in Asian Indians in North India. J Diabetes Complications 29(1):38–43. doi:10.1016/j.jdiacomp.2014.08.001

44. Nordstrand N, Gjevestad E, Dinh KN, Hofso D, Roislien J, Saltvedt E, Os I, Hjelmesaeth J (2011) The relationship between various measures of obesity and arterial stiffness in morbidly obese patients. BMC Cardiovasc Disord 11:7. doi:10.1186/1471-2261-11-7

45. Wykretowicz A, Adamska K, Guzik P, Krauze T, Wysocki H (2007) Indices of vascular stiffness and wave reflection in relation to body mass index or body fat in healthy subjects. Clin Exp Pharmacol Physiol 34(10):1005–1009. doi:10.1111/j.1440-1681.2007.04666.x

46. Corden B, Keenan NG, de Marvao AS, Dawes TJ, Decesare A, Diamond T, Durighel G, Hughes AD, Cook SA, O'Regan DP (2013) Body fat is associated with reduced aortic stiffness until middle age. Hypertension 61(6):1322–1327. doi:10.1161/hypertensionaha.113.01177

47. Czernichow S, Bertrais S, Oppert JM, Galan P, Blacher J, Ducimetiere P, Hercberg S, Zureik M (2005) Body composition and fat repartition in relation to structure and function of large arteries in middle-aged adults (the SU.VI.MAX study). Int J Obes (Lond) 29(7):826–832. doi:10.1038/sj.ijo.0802986

48. Sutton-Tyrrell K, Newman A, Simonsick EM, Havlik R, Pahor M, Lakatta E, Spurgeon H, Vaitkevicius P (2001) Aortic stiffness is associated with visceral adiposity in older adults enrolled in the study of health, aging, and body composition. Hypertension 38(3):429–433

49. Benetos A, Zervoudaki A, Kearney-Schwartz A, Perret-Guillaume C, Pascal-Vigneron V, Lacolley P, Labat C, Weryha G (2009) Effects of lean and fat mass on bone mineral density and arterial stiffness in elderly men. Osteoporos Int 20(8):1385–1391. doi:10.1007/s00198-008-0807-8

50. Fantin F, Rossi AP, Cazzadori M, Comellato G, Mazzali G, Gozzoli MP, Grison E, Zamboni M (2013) Central and peripheral fat and subclinical vascular damage in older women. Age Ageing 42(3):359–365. doi:10.1093/ageing/aft005

51. Tanaka H, DeSouza CA, Seals DR (1998) Absence of age-related increase in central arterial stiffness in physically active women. Arterioscler Thromb Vasc Biol 18(1):127–132

52. Zhu W, Hooker SP, Sun Y, Xie M, Su H, Cao J (2014) Associations of cardiorespiratory fitness with cardiovascular disease risk factors in middle-aged Chinese women: a cross-sectional study. BMC Womens Health 14:62. doi:10.1186/1472-6874-14-62

53. Quan HL, Blizzard CL, Sharman JE, Magnussen CG, Dwyer T, Raitakari O, Cheung M, Venn AJ (2014) Resting heart rate and the association of physical fitness with carotid artery stiffness. Am J Hypertens 27(1):65–71. doi:10.1093/ajh/hpt161

54. Otsuki T, Maeda S, Iemitsu M, Saito Y, Tanimura Y, Ajisaka R, Miyauchi T (2007) Vascular endothelium-derived factors and arterial stiffness in strength- and endurance-trained men. Am J Physiol Heart Circ Physiol 292(2):H786–H791. doi:10.1152/ajpheart.00678.2006

55. Tarumi T, Gonzales MM, Fallow B, Nualnim N, Pyron M, Tanaka H, Haley AP (2013) Central artery stiffness, neuropsychological function, and cerebral perfusion in sedentary and endurance-trained middle-aged adults. J Hypertens 31(12):2400–2409. doi:10.1097/HJH.0b013e328364decc

56. Bertovic DA, Waddell TK, Gatzka CD, Cameron JD, Dart AM, Kingwell BA (1999) Muscular strength training is associated with low arterial compliance and high pulse pressure. Hypertension 33(6):1385–1391

57. Miyachi M, Kawano H, Sugawara J, Takahashi K, Hayashi K, Yamazaki K, Tabata I, Tanaka H (2004) Unfavorable effects of resistance training on central arterial compliance: a randomized intervention study. Circulation 110(18):2858–2863. doi:10.1161/01.cir.0000146380.08401.99

58. Kawano H, Tanimoto M, Yamamoto K, Sanada K, Gando Y, Tabata I, Higuchi M, Miyachi M (2008) Resistance training in men is associated with increased arterial stiffness and blood pressure but does not adversely affect endothelial function as measured by arterial reactivity to the cold pressor test. Exp Physiol 93(2):296–302. doi:10.1113/expphysiol.2007.039867

59. Cook JN, DeVan AE, Schleifer JL, Anton MM, Cortez-Cooper MY, Tanaka H (2006) Arterial compliance of rowers: implications for combined aerobic and strength training on arterial elasticity. Am J Physiol Heart Circ Physiol 290(4):H1596–H1600. doi:10.1152/ajpheart.01054.2005

60. Mero AA, Hulmi JJ, Salmijarvi H, Katajavuori M, Haverinen M, Holviala J, Ridanpaa T, Hakkinen K, Kovanen V, Ahtiainen JP, Selanne H (2013) Resistance training induced increase in muscle fiber size in young and older men. Eur J Appl Physiol 113(3):641–650. doi:10.1007/s00421-012-2466-x

61. Mitchell GF (2015) Arterial stiffness: insights from Framingham and Iceland. Curr Opin Nephrol Hypertens 24(1):1–7. doi:10.1097/mnh.0000000000000092

62. Park JB, Schiffrin EL (2001) Small artery remodeling is the most prevalent (earliest?) form of target organ damage in mild essential hypertension. J Hypertens 19(5):921–930

63. Palmieri V, Celentano A, Roman MJ, de Simone G, Lewis MR, Best L, Lee ET, Robbins DC, Howard BV, Devereux RB (2001) Fibrinogen and preclinical echocardiographic target organ damage: the strong heart study. Hypertension 38(5):1068–1074

64. Desai AS, Mitchell GF, Fang JC, Creager MA (2009) Central aortic stiffness is increased in patients with heart failure and preserved ejection fraction. J Card Fail 15(8):658–664. doi:10.1016/j.cardfail.2009.03.006

65. Celik T, Yuksel UC, Fici F, Celik M, Yaman H, Kilic S, Iyisoy A, Dell'oro R, Grassi G, Yokusoglu M, Mancia G (2013) Vascular inflammation and aortic stiffness relate to early left ventricular diastolic dysfunction in prehypertension. Blood Press 22(2):94–100. doi:10.3109/08037051.2012.716580

66. Totaro S, Khoury PR, Kimball TR, Dolan LM, Urbina EM (2015) Arterial stiffness is increased in young normotensive subjects with high central blood pressure. J Am Soc Hypertens 9(4):285–292. doi:10.1016/j.jash.2015.01.013

67. Coutinho T, Turner ST, Kullo IJ (2011) Aortic pulse wave velocity is associated with measures of subclinical target organ damage. JACC Cardiovasc Imaging 4(7):754–761. doi:10.1016/j.jcmg.2011.04.011

68. Mitchell GF, van Buchem MA, Sigurdsson S, Gotal JD, Jonsdottir MK, Kjartansson O, Garcia M, Aspelund T, Harris TB, Gudnason V, Launer LJ (2011) Arterial stiffness, pressure and flow pulsatility and brain structure and function: the Age, Gene/Environment Susceptibility—Reykjavik study. Brain 134(pt 11):3398–3407. doi:10.1093/brain/awr253

69. Kidher E, Patel VM, Nihoyannopoulos P, Anderson JR, Chukwuemeka A, Francis DP, Ashrafian H, Athanasiou T (2014) Aortic stiffness is related to the ischemic brain injury biomarker N-methyl-D-aspartate receptor antibody levels in aortic valve replacement. Neurol Res Int 2014:970793. doi:10.1155/2014/970793

70. Brandts A, van Elderen SG, Westenberg JJ, van der Grond J, van Buchem MA, Huisman MV, Kroft LJ, Tamsma JT, de Roos A (2009) Association of aortic arch pulse wave velocity with left ventricular mass and lacunar brain infarcts in hypertensive patients: assessment with MR imaging. Radiology 253(3):681–688. doi:10.1148/radiol.2533082264

71. Jia G, Aroor AR, Sowers JR (2014) Arterial stiffness: a nexus between cardiac and renal disease. Cardiorenal Med 4(1):60–71. doi:10.1159/000360867

72. Smith A, Karalliedde J, De Angelis L, Goldsmith D, Viberti G (2005) Aortic pulse wave velocity and albuminuria in patients with type 2 diabetes. J Am Soc Nephrol 16(4):1069–1075. doi:10.1681/asn.2004090769

73. Townsend RR, Wimmer NJ, Chirinos JA, Parsa A, Weir M, Perumal K, Lash JP, Chen J, Steigerwalt SP, Flack J, Go AS, Rafey M, Rahman M, Sheridan A, Gadegbeku CA, Robinson NA, Joffe M (2010) Aortic PWV in chronic kidney disease: a CRIC ancillary study. Am J Hypertens 23(3):282–289. doi:10.1038/ajh.2009.240

74. Yokoyama H, Aoki T, Imahori M, Kuramitsu M (2004) Subclinical atherosclerosis is increased in type 2 diabetic patients with microalbuminuria evaluated by intima-media thickness and pulse wave velocity. Kidney Int 66(1):448–454. doi:10.1111/j.1523-1755.2004.00752.x

75. Safar ME, London GM, Plante GE (2004) Arterial stiffness and kidney function. Hypertension 43(2):163–168. doi:10.1161/01.HYP.0000114571.75762.b0

Chapter 4
Interventions to Destiffen Arteries

Abstract Stiffening of arteries is a condition associated with aging and disease that results in a less compliant arterial system. Arterial stiffness is an important CVD risk factor that is associated with premature mortality and heightened morbidity. Intervening to reduce arterial stiffening is a challenge, yet there is mounting evidence supporting the effects of lifestyle and dietary modifications to reduce aortic pulse wave velocity (aPWV), the gold standard measure of aortic stiffness. Plant-based polyphenolic compounds, omega-3 fatty acids, antioxidant vitamins C and E, regular aerobic exercise, sodium reduction and weight loss have all shown to be successful strategies to destiffen arteries. First-line strategies to combat CVD risk factors such as arterial stiffening should include lifestyle modifications.

Keywords Nutraceutical • Polyphenols • Omega-3 • Sodium • Aerobic • Weight loss

Arterial Stiffness Overview

The detrimental effects of increased arterial stiffness are multifaceted, which include: (1) Increases in central systolic blood pressure; (2) Increase in left ventricular afterload; and (3) Decrease in diastolic perfusion pressure. Primarily due to these physiological mechanisms, stiffening of the aorta is an independent risk factor for cardiovascular disease (CVD) that is associated with heightened morbidity and, premature mortality when assessed by aortic pulse wave velocity (aPWV) [1]. Indeed arterial stiffness is a novel risk factor that predicts incident CVD, and prospective data supporting the proposed pathophysiological mechanisms and their deleterious consequences on cardiovascular health, is mounting, and convincing. Moreover, arterial stiffness accelerates target organ damage to the brain, heart, and kidneys. Thus, identifying interventions to reduce arterial stiffness, and ultimately CVD-related events is clinically relevant.

The intent of this chapter is to focus on interventions that destiffen arteries, specifically the aorta (Fig. 4.1). Both preventing and ameliorating arterial stiffness are of great importance so both will be reviewed. Because mechanistic insight for interventions is important, we will also summarize the overarching mechanisms that underlie these novel interventions. However, in order to clearly and concisely

© Springer International Publishing Switzerland 2015

B.S. Fleenor, A.J. Berrones, *Arterial Stiffness*, SpringerBriefs in Physiology,

DOI 10.1007/978-3-319-24844-8_4

Fig. 4.1 Reducing arterial
stiffness through functional
foods, exercise, and weight
loss

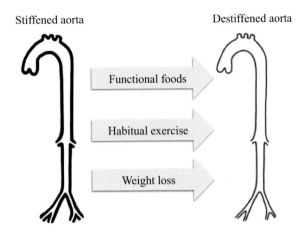

Stiffened aorta   Destiffened aorta

Functional foods

Habitual exercise

Weight loss

present this body of literature we have narrowed down our discussion of novel interventions to those shown to destiffen arteries that we believe have greater potential. As such, we have chosen to highlight interventions that meet the following criteria: (1) The intervention has to be equal to or longer than 4 weeks ($\geq$4 weeks); (2) The intervention must use carotid-to-femoral, aortic pulse wave velocity (aPWV) as its measure of arterial stiffness, which is the gold standard measure of aortic stiffness; (3) The intervention must have studied an adult ($\geq$18 years) human population; and (4) The intervention cannot use pharmaceutical or prescription-based medications as therapy to arterial stiffness.

Nutraceuticals and Functional Foods

The role of dietary factors in both the acceleration and slowing of CVD is undeniable. Skewing the energy balance equation towards greater intake vs. expenditure is the principal cause of obesity, which in turn accelerates arterial stiffening and the risk of CVD events. Understanding the impact of food intake, in terms of quality and quantity, and how it relates to CVD is a twenty-first century puzzle that remains to be solved. Many so-called "heart healthy" diets that have been conventionally recommended such as reduced saturated fat intake have shown poor associations with overall CV health status, thus complicating matters. It now appears as though the relation of saturated fat intake to CVD risk is dependent upon the *source*, be it plant or animal based [2, 3]. To date there is no consensus as to what exactly a "heart healthy" diet should consist of, and if in fact saturated fat intake is as pernicious as once thought.

The term "nutraceutical" is a portmanteau of the words "nutrition" and "pharmaceutical," and refers to any substance that is a food, or part of a food, that provides medical or health benefits including the prevention and treatment of disease [4]. Similarly, the definition of a "functional food" is any food or food ingredient that may provide a health benefit beyond the traditional nutrients it contains [4]. Because of the quasi-identical definitions and comparable interpretation, we have decided to use both terms interchangeably. We have defined, and will focus on three broad categories of functional foods that have been shown to have beneficial effects for attenuating arterial stiffness, which include: polyphenolic compounds, omega-3 fatty acids, and micronutrients.

It is worthwhile to mention that regular ingestion of functional foods have also been shown to positively impact hyperlipidemia by reducing serum lipid profiles through reductions in total cholesterol, triglycerides, LDL cholesterol, and elevations in HDL cholesterol [5]. Additionally, functional foods may boost cellular antioxidant capacities and thwart aging associated diseases such as cancer and neurological disorders [6]. Last, functional foods may alleviate symptoms of gut malfunction by bolstering anti-inflammatory cytokines that positively impact the gut microflora during allergic responses and infections seen with Crohn's disease [7]. Thus, it is reasonable to expect these beneficial effects of functional foods to be extended to arterial stiffness.

Polyphenolic Compounds

Polyphenols are classified on the basis of the number of phenol rings that they contain and by the structural elements that bind these rings to one another. Broadly, they may be divided into four major classes, including: (A) Phenolic acids, (B) Flavonoids, (C) Stilbenes, and (D) Lignans [8]. There are more than 4000 varieties of flavonoids that include six subclasses, of which isoflavones and anthocyanins will be focused on due to the available scientific evidence of these compounds to promote arterial destiffening.

Flavonoids

Epidemiological studies suggest that high intakes of dietary flavonoids are associated with decreased CVD mortality and risk factors [9]. Fruits and vegetables having a high concentration of flavonoids that are generally recognized as healthy include apples, berries, onions, tea leaves, and dark chocolate. A clinical trial using dark chocolate (10/g day; >75 % cocoa) for 4 weeks in healthy young adults (40 females; 20 males; mean age $= 20 \pm 2$ years) showed a significant decrease in aPWV compared to the control group (pre: 6.13 ± 0.41, post: 5.83 ± 0.53 m/s) [10]. These improvements in arterial stiffness were independent of peripheral (brachial) and central (aortic) blood pressures, which did not change throughout the 4-week period for the intervention group. However, an increase in flow-mediated dilation was increased with dark chocolate supplementation (13.91 ± 4.71 to 23.22 ± 7.64 %), suggesting this key functional aspect of arterial stiffness contributes to the destiffening process. The proposed mechanism for decreased arterial stiffness and enhanced endothelial function in this dark chocolate intervention was increased plasma epicatechin, which enhances NO production and bioavailability leading to increased endothelium-dependent vasodilation [11]. Thus, the cardioprotective effects of dark chocolate may be ascribed to cocoa flavonoids via improvements in NO bioavailability and arterial dilation.

Isoflavones

Isoflavones, which are components of soy protein, promote in vivo endothelium-dependent dilation and inhibit constrictor responses to collagen either by inhibiting platelet aggregation or platelet release of vasoconstrictors, or both [12]. However, clinical trials using isoflavone supplementation as a means to enhance arterial function have shown equivocal results. Only two original reports that have shown successful reductions in aortic stiffness from isoflavone administration will be discussed.

A double-blind, crossover design trial in healthy middle-aged normotensive adults examined the effects of isoflavones enriched with biochanin or formonentin (metabolites of isoflavones; 80 mg/day) or placebo for 6-weeks [13]. While supplementation with isoflavones did not improve flow mediated dilation or blood pressure, aPWV was significantly decreased after the intervention compared with the placebo (pre: 8.69 ± 0.2, post: 8.39 ± 0.2 m/s). Additionally, total peripheral resistance was decreased (17.0 ± 0.7 to 16.0 ± 0.6 RU, $P=0.03$) and circulating pro-inflammatory molecule VCAM-1 was reduced by 11 % in the intervention compared with the placebo group following isoflavone supplementation ($P=0.009$). Although endothelial function did not directly improve following isoflavone ingestion for 6 weeks, aortic stiffness improved likely to due altered vascular tone rather than to structural changes [13].

Further evidence for isoflavones to improve arterial function was observed in a randomized, double-blind, crossover designed trial assessing the isoflavone metabolite trans-tetrahydrodaidzein (THD, a metabolite of daidzein; 1 g/day) for 5 weeks [14]. This study randomized 25 middle-aged overweight/obese subjects (14 overweight males; 11 overweight postmenopausal women; mean BMI $=30.3 \pm 4.7$; mean age $=57 \pm 7$ years) to THD or placebo, and aPWV was significantly decreased by 1.1 m/s following isoflavone supplementation, representing a 10 % improvement in aortic stiffness (placebo: 9.90, baseline intervention: 9.72, isoflavone supplementation: 8.8 m/s; $P<0.05$). Moreover, both systolic and diastolic blood pressures decreased with isoflavone treatment and differed significantly between the active treatment and placebo periods ($P<0.05$). Because arterial stiffness is generally related to blood pressure [15], the absence of correlation between the changes in pressure and aPWV suggests that isoflavone supplementation affected each of those two parameters independently [14].

Pigments (Anthocyanins and Betalains)

Consumption of fruits and vegetables is associated with a reduced risk of heart disease and premature mortality [16]. Pigments are compounds found in fruits and vegetables that have unique spectral properties, giving off an assortment of bright and colorful visual characteristics of wavelengths between 380 and 730 nm [17]. The three main classes of pigments for coloration in plants are (A) Anthocyanins, (B) Betalains, and (C) Carotenoids [17]. Cranberries, for example, have 678 mg/100 g of total phenols, and high levels of anthocyanins, which ranks them just below blueberries in terms of the level of high quantity/quality total phenol antioxidant concentrations [16]. Moreover, beetroots, a major source of betalain pigments, have strong antioxidant properties and increased consumption is associated with reduced CVD risk [18]. It has been hypothesized that cranberry and beetroot consumption impart specific cardiovascular benefits by reducing inflammation and improving lipid profiles.

In a randomized trial, older adults (30 males; 14 females; mean age = 62 ± 10 years) with coronary artery disease were given either cranberry juice (480 mL "double-strength" juice; 835 mg total polyphenols; 94 mg anthocyanins) or placebo juice for 4 weeks, in a crossover design [19]. aPWV following cranberry juice treatment was 7.8±2.2 m/s which was significantly reduced compared to baseline value (8.3±2.3 m/s; P=0.003) with no changes in the placebo group. There were no significant improvements in brachial artery flow mediated dilation, markers of inflammation, or blood pressure. However, a modest yet significant reduction of HDL cholesterol by 2 % (pre: 43±10 vs. post: 42±9 mg/dL; P=0.04) was discovered, which was unexpected and in contrast to other cranberry interventions.

Increasing NO bioavailability by consuming foods high in nitrates/nitrites has been identified as a means to reduce blood pressure. A randomized, placebo-controlled trial in hypertensive patients (26 males; 38 females; mean age = 57 ± 15 years) with hypertension given dietary nitrate (250 mL/day beetroot juice, a major source of betalains) had reductions in both systolic and diastolic blood pressure (P=0.05) [20]. Compared to baseline, beetroot juice treatment decreased aPWV by 0.59 m/s (P<0.01), and increased peak flow mediated dilation by 1.0 % (P<0.001). Thus, by capitalizing on NO metabolism, once daily dietary nitrate interventions may be practical and a cost-effective means to augment NO bioavailability, and reduce blood pressure and arterial stiffness in hypertensive patients [20].

Omega-3 Fatty Acids

Over the last 50 years studies have shown an array of health benefits among those following Mediterranean diets, including reduced CVD risk [21]. Several advantages of following a Mediterranean diet is improved vasodilation, decreased blood pressure, and reductions in oxidative stress and inflammatory biomarkers [22]. The purported mechanism of reduced risk of CVD by the Mediterranean diet is due to the greater intake of long chain polyunsaturated omega-3 fatty acids. The plant based food sources of omega-3 fatty acids (alpha linolenic acid [ALA]) include flaxseed, hempseed, soybeans, and walnuts [22, 23]. The marine sources of omega-3 fatty acids (eicosapentaenoic acid [EPA] and docosapentaenoic acid [DHA]) include algae, salmon, mackerel, and trout. Importantly, the metabolic conversion of the 18 carbon fatty acid ALA to the 20 carbon fatty acid EPA and 22 carbon fatty acid DHA is approximately 5 % in humans, thus making it "essential" to consume EPA and DHA in copious amounts [24]. In short, specific consideration should be given to the type of long chain polyunsaturated omega-3 fatty acid consumed, in particular EPA and DHA, due to their association with decreased CVD risk factors and improved vascular outcomes [25].

Unfortunately, the metabolic syndrome is a common disease, characterized by at least three of the following risk factors: abdominal obesity, low HDL cholesterol, high triglycerides, hypertension, and impaired fasting glucose. The metabolic syndrome accounts for up to 1/3rd of CVD in men [26], therefore assessing the influence of omega-3 fatty acids in a metabolic syndrome population is a promising therapy. As such, 12 weeks of oral supplementation with 2 g/day of omega-3 polyunsaturated fatty acids (46 % EPA; 38 % DHA) in middle-aged adults with metabolic syndrome reduced aPWV from 7.62 ± 1.59 at baseline to 7.22 ± 1.54 m/s ($P < 0.001$) at the end of the trial. Additionally, total cholesterol, triglycerides, LDL cholesterol, and IL-6 were significantly reduced, while plasminogen activator inhibitor 1 and flow mediated dilation increased, over the course of the 12-week intervention compared to baseline values ($P < 0.001$). The favorable effects of omega-3 supplementation on endothelial function and arterial stiffness in adults with metabolic syndrome appear to be partially mediated through modification of the metabolic profile and inflammatory processes [27].

As mentioned previously, endothelial dysfunction is an early event in the pathogenesis of arteriosclerosis that increases CVD risk. Cigarette smoking is associated with dose related and (potentially) reversible impairment of endothelium-dependent arterial dilation in asymptomatic young adults [28]. The smoking related impairment in endothelium-dependent arterial dilation promotes increases in the pro-inflammatory phenotype, leading to changes in the thrombosis/fibrinolysis system, which are important pathways for coagulation and the enzymatic breakdown of blood clots [29]. In a randomized, placebo-controlled, double-blind, and crossover study, the effects of 12-week omega-3 polyunsaturated fatty acids supplementation (46 % EPA; 38 % DHA) in 20 healthy smokers (13 males; 7 females; mean age = 28 ± 3 years) who had been smoking greater than 20 cigarettes per day for greater than 5 years were determined [30]. Treatment with omega-3 polyunsaturated fatty acids reduced aPWV from 5.87 ± 0.63 at baseline to 5.54 ± 0.76 m/s ($P = 0.007$) following the intervention. Additionally, TNF alpha and IL-6 were significantly reduced, while plasminogen activator inhibitor 1 and flow mediated dilation increased, over the course of the 12-week intervention compared with baseline values ($P < 0.05$). Supplementation with omega-3 fatty acids did not, however, influence the lipid profile. In short, the detrimental effects of regular cigarette smoking on endothelial function and elastic properties of the arterial tree can be ameliorated by long chain omega-3 fatty acids such as EPA and DHA in asymptomatic adults, in part, through reductions in inflammation [30].

Micronutrients

Sodium Restriction

The consumption of highly salted food is increasing worldwide, and these changes in salt intake present a major challenge on the physiological systems of the body [31]. Much evidence exists regarding our high salt intake and its role in creating the large rise in blood pressure that occurs with age [31]. Therefore, minimizing sodium intake has been an important strategy for addressing the age-associated increase in blood pressure, which is directly linked to the pathophysiological processes that underpin arterial stiffness. For example, Dietary Approaches to Stop Hypertension (DASH) have been successful in reducing systolic and diastolic blood pressures in a variety of adult populations, and so DASH-like diets that invoke these beneficial effects of reducing blood pressure and the concomitant decrease in arterial stiffness are highlighted in this subsection [32].

In a randomized, placebo-controlled study with a crossover design, eight men and three women (mean age = 60.2 years) with moderately elevated systolic blood pressure ($139\pm2/83\pm2$ mmHg) followed a low sodium (77 ± 99 mmol/day) or normal sodium (144 ± 7 mmol/day) diet for 5 weeks [33]. aPWV for the low sodium condition was significantly reduced compared to that of the normal sodium condition (7.0 ± 0.4 vs. 8.43 ± 0.36 m/s, P=0.001). Furthermore, systolic blood pressure was significantly reduced as a result of the low sodium condition compared with baseline and the normal sodium condition (low sodium: 127 ± 3, baseline: 139 ± 2, normal sodium: 138 ± 5, P<0.001). It is important to consider that systolic blood pressure is a key determinant of large elastic artery (aortic) stiffness [34], which may explain the impact of low sodium consumption on reductions of aPWV in this middle-aged and older cohort of adults with moderately elevated systolic blood pressure.

Similarly, a study in a postmenopausal cohort of women of middle and older age (n = 17; mean age = 65 ± 10 years) used a 3-month sodium restriction protocol (preintervention to postintervention: 2685 ± 559 vs. 1421 ± 512 mg/day) to assess changes in arterial function [35]. In addition to reductions in systolic blood pressure (~16 mmHg, P<0.05), aPWV was also significantly reduced as a result of 3 months of sodium restriction (8.7 ± 2.0 vs. ~7.3 ± 1.1 m/s, P<0.05). Again, this study highlights the influence of blood pressure reduction with sodium restriction to decrease central arterial stiffness in women whose systolic blood pressure ranged from 130 to 159 mmHg.

Additional studies have also demonstrated reductions in blood pressure and arterial stiffness in mildly hypertensive patients (n = 169; mean age = 50 ± 11 years; $147\pm13/91\pm8$ mmHg) as a result of modest sodium restriction [36]. In this study, the reduction in salt intake from 9.7 to 6.5 g/day was successful in reducing systolic blood pressure by 5 ± 12 mmHg (P<0.001). Importantly, aPWV also dropped from 11.5 ± 2.3 to 11.1 ± 1.9 m/s, which was statistically significant (P<0.01). These results demonstrate that a modest reduction in salt intake can improve CVD risk by attenuating systolic blood pressure and increasing compliance of the aorta. Because the reduction in salt intake was in accordance with current public health recommendations, the improvements (albeit slight) seen in arterial stiffness are promising not only for mildly hypertensive adults, but also normotensive adults. The research regarding sodium restriction in well-controlled clinical trials is undoubtedly scarce. However, the studies presented herein point to a meaningful connection between sodium consumption, systolic blood pressure, and arterial stiffness that deserves further clarification.

Antioxidant Vitamins (C and E)

It has been hypothesized that vascular function and structure will be improved if oxidative stress can be ameliorated by reducing free radicals and protecting nitric oxide (NO) from inactivation [37]. Supplementation with antioxidant vitamins C and E, in particular, are recognized to protect against lipid peroxidation, a contributor to arterial stiffness. And many observational studies have shown that a high dietary intake or high blood concentrations of antioxidant vitamins are associated with reduced risk of CVD in general [38]. Yet few studies have examined the potential effects of antioxidant supplementation on vascular stiffness and function in humans.

Using an emulsified preparation of a tocotrienol-rich vitamin E supplement, 36 healthy young men (mean age $= 23.9 \pm 0.39$ years) were assigned to either 50, 100, or 200 mg daily supplementation of this compound for 8 weeks [39]. All treatment groups had higher plasma concentrations of vitamin E following the 8-week study duration ($P < 0.05$). Vitamin E supplementation had no effect on systolic blood pressure. However, aPWV was significantly reduced in both the 100 and 200 mg supplementation groups by 0.77 and 0.65 m/s, respectively. This reduction in aortic stiffness was approximately 10 % less than the values observed at baseline. Mechanistically, it has been shown that free radicals inactivate endothelium-derived relaxing factor (EDRF), an important vasodilator [40]. Therefore, it is plausible that the reductions in aPWV seen in this study were due to neutralization of free radicals, and preserved EDRF activity.

Vitamin C supplementation has been shown to restore NO activity by improving endothelium-dependent vasodilation in adult patients with essential hypertension [41]. The synergistic effects of vitamin C and E supplementation may exert a greater antioxidant effect beyond those effects in isolation. In a randomized, double-blind, placebo-controlled study with a crossover design a combined vitamin C (1 g) and vitamin E (400 IU) treatment was given for 8 weeks to 30 males (mean age $= 50$ years; range 42–60) with essential hypertension (never treated) [42]. While combined treatment of vitamin C and vitamin E did not affect blood pressure, a significant reduction in aPWV was observed (pre to post, ~9.1 to ~8.4 m/s, $P < 0.01$). The combined effect of antioxidant vitamins C together with E may provide optimal conditions for endothelial NO formation [43]. Therefore, independent of changes in blood pressure, improvements in aortic stiffness were demonstrated in this male population of untreated hypertensives that is possibly due to decreases in vascular oxidative stress, and enhanced NO bioavailability.

Chronic Exercise Training

Regular aerobic exercise is an important strategy to bolster cardiovascular function and prolong longevity. The benefits of aerobic exercise are proportional to intensity and duration, that is, the greater and more robust the exercise program (in general) then greater cardiovascular improvements will be observed. The gold standard measure of cardiorespiratory fitness is maximal oxygen uptake (VO_{2MAX}), expressed *absolutely* as liters of oxygen per minute (L/min) or *relatively* as milliliters of oxygen per kilogram of body mass per minute (mL/kg/min). Studies have shown that regular aerobic exercise inhibits large elastic artery stiffening with aging by optimizing aortic stiffening, which is associated with preserved endothelial function [44]. Therefore, engaging in regular aerobic exercise can be considered both a preventative strategy and treatment for age-associated arterial dysfunction, thus suggesting endothelial function is an important mechanism for overall improvements in arterial function.

The improvements in arterial stiffness in otherwise healthy adults is most likely due to enhanced NO synthesis and activity. As such, in an experimental animal model reductions in aortic stiffness with chronic aerobic exercise training were associated with increased endothelial nitric oxide synthase (eNOS), a critical enzyme for NO production stiffness [45]. Increased NO bioavailability may, in part, decrease vascular tone and blood pressure. Because of this sustained reduction in vascular smooth muscle tone, the structural composition of the arteries would slowly change, as the arteries would remodel to meet the demands of reduced intra-arterial pressure brought on by aerobic exercise. If the arterial pressure were consistently reduced due to greater eNOS activation and NO synthesis, the extracellular matrix (ECM) protein profile would favorably adjust to further contribute to the **destiffening** process.

Additional studies have also demonstrated beneficial effects of chronic aerobic exercise to destiffen arteries. However, these studies provide less mechanistic insight. In 17 middle-aged sedentary men (mean age = 50 ± 3 years) subjects were progressed in a brisk walking/jogging exercise-training program over a 16-week period where the final intensity was equivalent to 75 % of HRR ($HRR = HR_{MAX} - HR_{REST}$) for 45 min/day, 3–4 days/week [46]. After the exercise intervention aPWV was significantly decreased from 9.37 ± 3.4 to 8.7 ± 3.2 m/s (P<0.05).

Using a similarly aged cohort (range age = 32 to 59 years) of women (N = 12), it was shown that 2 days/week for 30 min/day at 60–70 % of individual VO_{2MAX} for 12 weeks significantly reduced aPWV (P<0.05) [47]. Additionally, in a younger middle-aged group of sedentary adults (58 females; 19 males; range age = 30–55 years) who performed 6 months of accumulated (3 sessions × 10 min/ day, 5 days/week) brisk walking at a moderate intensity (measured with an accelerometer) also displayed reductions in aPWV [48]. In summary, these studies [46–48] demonstrate that reductions in arterial stiffness were independent of peripheral (brachial) blood pressure values, which did not change over the intervention period for any study.

The efficacy of chronic aerobic exercise training has also been assessed in populations with increased CVD risk. For instance, middle-aged patients (12 females; 4 males; mean age = 45 ± 9 years) with hypertension underwent interval exercise training (3 days/week; alternating intensity from 50 % × 2 min and 80 % × 1 min of HRR for 40 min on a treadmill) for 16 weeks. This exercise protocol reduced aPWV from 9.44 ± 0.91 to 8.90 ± 0.96 m/s (P=0.009) [49]. Similarly, stage 1 hypertensives (10 males; 5 females; mean age = 49.8 ± 1.6 years) performing aerobic exercise (3 days/

week for 4 weeks of 30 min of treadmill exercise at 65 % VO_{2MAX}) found a significant decrease in aPWV by 9.5 % (P<0.05) and a significant reduction in mean arterial pressure (MAP, pre vs. post: 103.5 ± 1.8 vs. 100.3 ± 2 mmHg, P<0.05) [50]. Significant reductions in aPWV have also been observed in older subjects who had co-morbidities such as type 2 diabetes, hypertension, hyperglycemia, and hypercholesterolemia after 3 months of vigorous regular aerobic exercise [51, 52].

It should also be noted that in apparently healthy young adults who engage in high intensity resistance training demonstrate increased arterial stiffening [53]. Importantly, combined aerobic and resistance training may have reduced beneficial effects on arterial stiffness compared with chronic aerobic exercise interventions alone [54]. These data do not suggest that regular resistance training is detrimental to overall cardiovascular health. In fact, combined aerobic and resistance training is most effective in the chronic modification of blood pressure and lipid profile, and in the reduction of total CVD risk factors aggregated [55].

Lifestyle Modifications

The first-line approach for prevention and treatment of most CVD risk factors is lifestyle & behavior modifications [56]. While regular aerobic exercise is beneficial and promotes arterial compliance, it is possible to partake in an exercise program and lose little, if any, body weight. The impact of excessive body mass on arterial stiffness is clear: The larger you are, the stiffer your arteries, independent of age [57]. Thus, the physiological basis of weight gain (or weight loss) ultimately points back to the law of conservation of energy, which states that energy cannot be created or destroyed but rather changes forms. Because food provides the human body with energy, and is accumulated then subsequently stored in adipose tissue, obesity can best be explained as a condition of excessive caloric intake relative to caloric expenditure. The health benefits of a lifestyle modification such as weight loss can therefore be achieved either by (1) eating less, or (2) increasing levels of physical activity.

Weight Loss

Among normotensive overweight/obese young adults, weight loss is associated with a reduction in aortic stiffness [58]. After a 1-year lifestyle intervention (diet/physical activity/reduced sodium), aPWV significantly decreased by 0.581 m/s after 6 months (P<0.0001), by 0.32 m/s (P=0.02) and after 12 months [58]. This 1-year intervention resulted in an average weight loss of 6.4 % (P<0.05). Additionally, systolic blood pressure, triglycerides, CRP, insulin, and leptin decreased (P<0.05); whereas, HDL cholesterol increased (P<0.05).

Additional studies have also demonstrated the effectiveness of lifestyle intervention on arterial stiffness in populations with greater CVD risk. Obese men (mean age = 45 ± 2 years; BMI = 30 ± 1 kg/m^2) were placed on a low-calorie lifestyle intervention (1380 kcal/day) for 12 weeks. aPWV significantly decreased from 9.79 ± 0.45 to 9.18 ± 0.29 m/s ($P < 0.05$). Additionally, plasma endothelin-1 (a vaso-constrictor) significantly decreased from 1.9 ± 0.1 to 1.3 ± 0.3 pg/mL ($P < 0.01$), and plasma nitrite/nitrate significantly increased from 24 ± 3 to 39 ± 4 µmol/L ($P < 0.01$), suggesting that enhanced endothelial function is responsible for the reduction (9 %, $P < 0.01$) in aortic stiffness following weight loss due to a low-calorie diet. Other physiological variables that changed over the course of the 12-week weight loss intervention were total cholesterol, triglycerides, and blood glucose ($P < 0.05$) [59].

Furthermore, a 12-week hypocaloric diet (1200 to 1500 kcal/day) administered to 25 overweight/obese, middle-aged and older adults (16 females; 9 males; mean age = 61.2 ± 0.8 years; BMI = 30.0 ± 0.6) demonstrated a significant reduction in aPWV by 1.87 ± 0.29 m/s ($P < 0.05$) [60]. The magnitude of this improvement in arterial stiffness was positively related to the percent weight loss ($r = 0.59$; $P < 0.05$). Additionally, brachial and aortic systolic blood pressures, triglycerides, total cholesterol, and glucose were significantly reduced ($P < 0.05$). Notably, the observed reductions in arterial stiffness in this study (~1.5 to 2.0 m/s) would translate into a reversal of age-related arterial stiffening by ~15 to 20 years [60].

Summary

The prevention of arterial stiffness is of paramount importance. Although arterial stiffness is largely a process of aging, obesity early in life can accelerate arterial stiffness resulting in development of premature CVD. Current guidelines recommend preventing and treating most CVD risk factors with 3–6 months of lifestyle interventions, followed by pharmacological therapy (if necessary) in order to achieve normalization [56]. We purposely did not include short term (<4 weeks) or pharmaceutical interventions for this reason. Our goal is to emphasize the importance of undertaking a healthy lifestyle by incorporating high quality nutrients in the form of polyphenolic compounds (e.g., flavonoids, isoflavones, pigments), omega-3 polyunsaturated fatty acids, reduced sodium intake, antioxidant vitamins C and E, regular aerobic exercise, and minimal consumption of calories needed to sustain adequate metabolic function. In short, an appropriate lifestyle is the first approach for achieving healthy arterial function that can ultimately reduce CVD risk.

References

1. Laurent S, Cockcroft J, Van Bortel L, Boutouyrie P, Giannattasio C, Hayoz D, Pannier B, Vlachopoulos C, Wilkinson I, Struijker-Boudier H (2006) Expert consensus document on arterial stiffness: methodological issues and clinical applications. Eur Heart J 27(21):2588–2605. doi:10.1093/eurheartj/ehl254
2. Babu AS, Veluswamy SK, Arena R, Guazzi M, Lavie CJ (2014) Virgin coconut oil and its potential cardioprotective effects. Postgrad Med 126(7):76–83. doi:10.3810/pgm.2014.11.2835

3. de Oliveira Otto MC, Mozaffarian D, Kromhout D, Bertoni AG, Sibley CT, Jacobs DR Jr, Nettleton JA (2012) Dietary intake of saturated fat by food source and incident cardiovascular disease: the Multi-Ethnic Study of Atherosclerosis. Am J Clin Nutr 96(2):397–404. doi:10.3945/ajcn.112.037770
4. Alissa EM, Ferns GA (2012) Functional foods and nutraceuticals in the primary prevention of cardiovascular diseases. J Nutr Metab 2012:569486. doi:10.1155/2012/569486
5. Chen G, Wang H, Zhang X, Yang ST (2014) Nutraceuticals and functional foods in the management of hyperlipidemia. Crit Rev Food Sci Nutr 54(9):1180–1201. doi:10.1080/10408 398.2011.629354
6. Ferrari CK, Torres EA (2003) Biochemical pharmacology of functional foods and prevention of chronic diseases of aging. Biomed Pharmacother 57(5–6):251–260
7. Cencic A, Chingwaru W (2010) The role of functional foods, nutraceuticals, and food supplements in intestinal health. Nutrients 2(6):611–625. doi:10.3390/nu2060611
8. Pandey KB, Rizvi SI (2009) Plant polyphenols as dietary antioxidants in human health and disease. Oxid Med Cell Longev 2(5):270–278. doi:10.4161/oxim.2.5.9498
9. Toh JY, Tan VM, Lim PC, Lim ST, Chong MF (2013) Flavonoids from fruit and vegetables: a focus on cardiovascular risk factors. Curr Atheroscler Rep 15(12):368. doi:10.1007/s11883 -013-0368-y
10. Pereira T, Maldonado J, Laranjeiro M, Coutinho R, Cardoso E, Andrade I, Conde J (2014) Central arterial hemodynamic effects of dark chocolate ingestion in young healthy people: a randomized and controlled trial. Cardiol Res Pract 2014:945951. doi:10.1155/2014/945951
11. Engler MB, Engler MM, Chen CY, Malloy MJ, Browne A, Chiu EY, Kwak HK, Milbury P, Paul SM, Blumberg J, Mietus-Snyder ML (2004) Flavonoid-rich dark chocolate improves endothelial function and increases plasma epicatechin concentrations in healthy adults. J Am Coll Nutr 23(3):197–204
12. Williams JK, Clarkson TB (1998) Dietary soy isoflavones inhibit in-vivo constrictor responses of coronary arteries to collagen-induced platelet activation. Coron Artery Dis 9(11):759–764
13. Teede HJ, McGrath BP, DeSilva L, Cehun M, Fassoulakis A, Nestel PJ (2003) Isoflavones reduce arterial stiffness: a placebo-controlled study in men and postmenopausal women. Arterioscler Thromb Vasc Biol 23(6):1066–1071. doi:10.1161/01.atv.0000072967.97296.4a
14. Nestel P, Fujii A, Zhang L (2007) An isoflavone metabolite reduces arterial stiffness and blood pressure in overweight men and postmenopausal women. Atherosclerosis 192(1):184–189. doi:10.1016/j.atherosclerosis.2006.04.033
15. Safar ME, Levy BI, Struijker-Boudier H (2003) Current perspectives on arterial stiffness and pulse pressure in hypertension and cardiovascular diseases. Circulation 107(22):2864–2869. doi:10.1161/01.cir.0000069826.36125.b4
16. Vinson JA, Su X, Zubik L, Bose P (2001) Phenol antioxidant quantity and quality in foods: fruits. J Agric Food Chem 49(11):5315–5321
17. Tanaka Y, Sasaki N, Ohmiya A (2008) Biosynthesis of plant pigments: anthocyanins, betalains and carotenoids. Plant J 54(4):733–749. doi:10.1111/j.1365-313X.2008.03447.x
18. Kazimierczak R, Hallmann E, Lipowski J, Drela N, Kowalik A, Pussa T, Matt D, Luik A, Gozdowski D, Rembialkowska E (2014) Beetroot (Beta vulgaris L.) and naturally fermented beetroot juices from organic and conventional production: metabolomics, antioxidant levels and anticancer activity. J Sci Food Agric 94(13):2618–2629. doi:10.1002/jsfa.6722
19. Dohadwala MM, Holbrook M, Hamburg NM, Shenouda SM, Chung WB, Titas M, Kluge MA, Wang N, Palmisano J, Milbury PE, Blumberg JB, Vita JA (2011) Effects of cranberry juice consumption on vascular function in patients with coronary artery disease. Am J Clin Nutr 93(5):934–940. doi:10.3945/ajcn.110.004242
20. Kapil V, Khambata RS, Robertson A, Caulfield MJ, Ahluwalia A (2015) Dietary nitrate provides sustained blood pressure lowering in hypertensive patients: a randomized, phase 2, double-blind, placebo-controlled study. Hypertension 65(2):320–327. doi:10.1161/hypertensionaha.114.04675
21. Scoditti E, Capurso C, Capurso A, Massaro M (2014) Vascular effects of the Mediterranean diet-part II: role of omega-3 fatty acids and olive oil polyphenols. Vascul Pharmacol 63(3):127–134. doi:10.1016/j.vph.2014.07.001

22. Vrablik M, Prusikova M, Snejdrlova M, Zlatohlavek L (2009) Omega-3 fatty acids and cardio-vascular disease risk: do we understand the relationship? Physiol Res 58(suppl 1):S19–S26
23. Sanders TA (2014) Plant compared with marine n-3 fatty acid effects on cardiovascular risk factors and outcomes: what is the verdict? Am J Clin Nutr 100(suppl 1):453S–458S. doi:10.3945/ajcn.113.071555
24. Brenna JT (2002) Efficiency of conversion of alpha-linolenic acid to long chain n-3 fatty acids in man. Curr Opin Clin Nutr Metab Care 5(2):127–132
25. Breslow JL (2006) n-3 fatty acids and cardiovascular disease. Am J Clin Nutr 83(6 suppl): 1477S–1482S
26. Wilson PW, D'Agostino RB, Parise H, Sullivan L, Meigs JB (2005) Metabolic syndrome as a precursor of cardiovascular disease and type 2 diabetes mellitus. Circulation 112(20):3066–3072. doi:10.1161/circulationaha.105.539528
27. Tousoulis D, Plastiras A, Siasos G, Oikonomou E, Verveniotis A, Kokkou E, Maniatis K, Gouliopoulos N, Miliou A, Paraskevopoulos T, Stefanadis C (2014) Omega-3 PUFAs improved endothelial function and arterial stiffness with a parallel antiinflammatory effect in adults with metabolic syndrome. Atherosclerosis 232(1):10–16. doi:10.1016/j.atherosclerosis. 2013.10.014
28. Celermajer DS, Sorensen KE, Georgakopoulos D, Bull C, Thomas O, Robinson J, Deanfield JE (1993) Cigarette smoking is associated with dose-related and potentially reversible impair-ment of endothelium-dependent dilation in healthy young adults. Circulation 88(5 pt 1): 2149–2155
29. Antoniades C, Tousoulis D, Vasiliadou C, Marinou K, Tentolouris C, Ntarladimas I, Stefanadis C (2004) Combined effects of smoking and hypercholesterolemia on inflammatory process, throm-bosis/fibrinolysis system, and forearm hyperemic response. Am J Cardiol 94(9):1181–1184. doi:10.1016/j.amjcard.2004.07.090
30. Siasos G, Tousoulis D, Oikonomou E, Zaromitidou M, Verveniotis A, Plastiras A, Kioufis S, Maniatis K, Miliou A, Siasou Z, Stefanadis C, Papavassiliou AG (2013) Effects of Omega-3 fatty acids on endothelial function, arterial wall properties, inflammatory and fibrinolytic status in smokers: a cross over study. Int J Cardiol 166(2):340–346. doi:10.1016/j.ijcard.2011.10.081
31. He FJ, MacGregor GA (2009) A comprehensive review on salt and health and current experi-ence of worldwide salt reduction programmes. J Hum Hypertens 23(6):363–384. doi:10.1038/jhh.2008.144
32. Saneei P, Salehi-Abargouei A, Esmaillzadeh A, Azadbakht L (2014) Influence of Dietary Approaches to Stop Hypertension (DASH) diet on blood pressure: a systematic review and meta-analysis on randomized controlled trials. Nutr Metab Cardiovasc Dis 24(12):1253–1261. doi:10.1016/j.numecd.2014.06.008
33. Jablonski KL, Fedorova OV, Racine ML, Geolfos CJ, Gates PE, Chonchol M, Fleenor BS, Lakatta EG, Bagrov AY, Seals DR (2013) Dietary sodium restriction and association with urinary marinobufagenin, blood pressure, and aortic stiffness. Clin J Am Soc Nephrol 8(11):1952–1959. doi:10.2215/cjn.00900113
34. Kaess BM, Rong J, Larson MG, Hamburg NM, Vita JA, Levy D, Benjamin EJ, Vasan RS, Mitchell GF (2012) Aortic stiffness, blood pressure progression, and incident hypertension. JAMA 308(9):875–881. doi:10.1001/2012.jama.10503
35. Seals DR, Tanaka H, Clevenger CM, Monahan KD, Reiling MJ, Hiatt WR, Davy KP, DeSouza CA (2001) Blood pressure reductions with exercise and sodium restriction in postmenopausal women with elevated systolic pressure: role of arterial stiffness. J Am Coll Cardiol 38(2):506–513
36. He FJ, Marciniak M, Visagie E, Markandu ND, Anand V, Dalton RN, MacGregor GA (2009) Effect of modest salt reduction on blood pressure, urinary albumin, and pulse wave velocity in white, black, and Asian mild hypertensives. Hypertension 54(3):482–488. doi:10.1161/hypertensionaha.109.133223
37. Leopold JA (2013) Cellular and molecular mechanisms of arterial stiffness associated with obesity. Hypertension 62(6):1003–1004. doi:10.1161/hypertensionaha.113.01872
38. Sesso HD, Buring JE, Christen WG, Kurth T, Belanger C, MacFadyen J, Bubes V, Manson JE, Glynn RJ, Gaziano JM (2008) Vitamins E and C in the prevention of cardiovascular disease in

men: the Physicians' Health Study II randomized controlled trial. JAMA 300(18):2123–2133. doi:10.1001/jama.2008.600

39. Rasool AH, Rahman AR, Yuen KH, Wong AR (2008) Arterial compliance and vitamin E blood levels with a self emulsifying preparation of tocotrienol rich vitamin E. Arch Pharm Res 31(9):1212–1217. doi:10.1007/s12272-001-1291-5

40. Rubanyi GM, Vanhoutte PM (1986) Superoxide anions and hyperoxia inactivate endothelium-derived relaxing factor. Am J Physiol 250(5 pt 2):H822–H827

41. Taddei S, Virdis A, Ghiadoni L, Magagna A, Salvetti A (1998) Vitamin C improves endothelium-dependent vasodilation by restoring nitric oxide activity in essential hypertension. Circulation 97(22):2222–2229

42. Plantinga Y, Ghiadoni L, Magagna A, Giannarelli C, Franzoni F, Taddei S, Salvetti A (2007) Supplementation with vitamins C and E improves arterial stiffness and endothelial function in essential hypertensive patients. Am J Hypertens 20(4):392–397. doi:10.1016/j.amjhyper. 2006.09.021

43. Heller R, Werner-Felmayer G, Werner ER (2004) Alpha-Tocopherol and endothelial nitric oxide synthesis. Ann N Y Acad Sci 1031:74–85. doi:10.1196/annals.1331.007

44. Seals DR (2014) Edward F. Adolph distinguished lecture: the remarkable anti-aging effects of aerobic exercise on systemic arteries. J Appl Physiol 117(5):425–439. doi:10.1152/japplphysiol.00362.2014

45. Maeda S, Iemitsu M, Miyauchi T, Kuno S, Matsuda M, Tanaka H (2005) Aortic stiffness and aerobic exercise: mechanistic insight from microarray analyses. Med Sci Sports Exerc 37(10): 1710–1716

46. Hayashi K, Sugawara J, Komine H, Maeda S, Yokoi T (2005) Effects of aerobic exercise training on the stiffness of central and peripheral arteries in middle-aged sedentary men. Jpn J Physiol 55(4):235–239. doi:10.2170/jjphysiol.S2116

47. Yoshizawa M, Maeda S, Miyaki A, Misono M, Saito Y, Tanabe K, Kuno S, Ajisaka R (2009) Effect of 12 weeks of moderate-intensity resistance training on arterial stiffness: a randomised controlled trial in women aged 32-59 years. Br J Sports Med 43(8):615–618. doi:10.1136/bjsm.2008.052126

48. Kearney TM, Murphy MH, Davison GW, O'Kane MJ, Gallagher AM (2014) Accumulated brisk walking reduces arterial stiffness in overweight adults: evidence from a randomized control trial. J Am Soc Hypertens 8(2):117–126. doi:10.1016/j.jash.2013.10.001

49. Guimaraes GV, Ciolac EG, Carvalho VO, D'Avila VM, Bortolotto LA, Bocchi EA (2010) Effects of continuous vs. interval exercise training on blood pressure and arterial stiffness in treated hypertension. Hypertens Res 33(6):627–632. doi:10.1038/hr.2010.42

50. Collier SR, Kanaley JA, Carhart R Jr, Frechette V, Tobin MM, Hall AK, Luckenbaugh AN, Fernhall B (2008) Effect of 4 weeks of aerobic or resistance exercise training on arterial stiffness, blood flow and blood pressure in pre- and stage-1 hypertensives. J Hum Hypertens 22(10):678–686. doi:10.1038/jhh.2008.36

51. Madden KM, Lockhart C, Cuff D, Potter TF, Meneilly GS (2009) Short-term aerobic exercise reduces arterial stiffness in older adults with type 2 diabetes, hypertension, and hypercholesterolemia. Diabetes Care 32(8):1531–1535. doi:10.2337/dc09-0149

52. Madden KM, Lockhart C, Cuff D, Potter TF, Meneilly GS (2013) Aerobic training-induced improvements in arterial stiffness are not sustained in older adults with multiple cardiovascular risk factors. J Hum Hypertens 27(5):335–339. doi:10.1038/jhh.2012.38

53. Miyachi M (2013) Effects of resistance training on arterial stiffness: a meta-analysis. Br J Sports Med 47(6):393–396. doi:10.1136/bjsports-2012-090488

54. Montero D, Vinet A, Roberts CK (2015) Effect of combined aerobic and resistance training versus aerobic training on arterial stiffness. Int J Cardiol 178:69–76. doi:10.1016/j.ijcard.2014.10.147

55. Sousa N, Mendes R, Abrantes C, Sampaio J, Oliveira J (2013) Long-term effects of aerobic training versus combined aerobic and resistance training in modifying cardiovascular disease risk factors in healthy elderly men. Geriatr Gerontol Int 13(4):928–935. doi:10.1111/ggi.12033

56. Tanaka H, Safar ME (2005) Influence of lifestyle modification on arterial stiffness and wave reflections. Am J Hypertens 18(1):137–144. doi:10.1016/j.amjhyper.2004.07.008

57. Scuteri A, Orru M, Morrell CH, Tarasov K, Schlessinger D, Uda M, Lakatta EG (2012) Associations of large artery structure and function with adiposity: effects of age, gender, and hypertension. The SardiNIA Study. Atherosclerosis 221(1):189–197. doi:10.1016/j.atherosclerosis.2011.11.045

58. Cooper JN, Buchanich JM, Youk A, Brooks MM, Barinas-Mitchell E, Conroy MB, Sutton-Tyrrell K (2012) Reductions in arterial stiffness with weight loss in overweight and obese young adults: potential mechanisms. Atherosclerosis 223(2):485–490. doi:10.1016/j.atherosclerosis.2012.05.022

59. Miyaki A, Maeda S, Yoshizawa M, Misono M, Saito Y, Sasai H, Endo T, Nakata Y, Tanaka K, Ajisaka R (2009) Effect of weight reduction with dietary intervention on arterial distensibility and endothelial function in obese men. Angiology 60(3):351–357. doi:10.1177/0003319708325449

60. Dengo AL, Dennis EA, Orr JS, Marinik EL, Ehrlich E, Davy BM, Davy KP (2010) Arterial destiffening with weight loss in overweight and obese middle-aged and older adults. Hypertension 55(4):855–861. doi:10.1161/hypertensionaha.109.147850

Index

© Springer International Publishing Switzerland 2015
B.S. Fleenor, A.J. Berrones, *Arterial Stiffness*, SpringerBriefs in Physiology,
DOI 10.1007/978-3-319-24844-8

Printed in the United States
By Bookmasters